Operator Theory: Advances and Applications
Volume 238

Founded in 1979 by Israel Gohberg

David E. Edmunds
W. Desmond Evans

Representations of Linear Operators Between Banach Spaces

 Birkhäuser

David E. Edmunds
Department of Mathematics
University of Sussex
Brighton
United Kingdom

W. Desmond Evans
School of Mathematics
University of Cardiff
Cardiff
United Kingdom

ISSN 0255-0156   ISSN 2296-4878 (electronic)
ISBN 978-3-0348-0641-1   ISBN 978-3-0348-0642-8 (eBook)
DOI 10.1007/978-3-0348-0642-8
Springer Basel Heidelberg New York Dordrecht London

Library of Congress Control Number: 2013948130

Mathematics Subject Classifications: 47A75, 47B06, 47B40, 35P30

Printed on acid-free paper

Springer Basel is part of Springer Science+Business Media (www.birkhauser-science.com)

Contents

Preface vii

Basic Notation xi

1 Preliminaries
  1.1 The geometry of Banach spaces 2
  1.2 Bases 21
  1.3 The p-trigonometric functions 42
  1.4 Entropy numbers and s-numbers 56
     1.4.1 Fundamentals 56
     1.4.2 Gelfand numbers and widths 60

2 Representation of Compact Linear Operators
  2.1 Compact operators in Hilbert spaces 67
  2.2 Compact operators in Banach spaces 71
     2.2.1 Preliminaries 71
     2.2.2 The linear projections S_k 79
     2.2.3 The nonlinear projections $P_k : X \to X_k$ 81
     2.2.4 The main convergence theorems 84
     2.2.5 A basis for X 90
     2.2.6 A Schmidt-type expansion for T 99
  2.3 Applications 104
     2.3.1 The p-Laplacian 105
     2.3.2 A weighted problem for the p-Laplacian 108
     2.3.3 A p-Laplacian problem in $\mathbb{R}^n$ 110
     2.3.4 The p-biharmonic operator 112
     2.3.5 Sturm–Liouville theory for the p-Laplacian 113
  2.4 Lusternik–Schnirelmann critical levels 115
     2.4.1 Comparison of eigenvalues 115
     2.4.2 A Hardy-type operator 118
  2.5 Further consequences of the boundedness of $(S_n)_{n\in\mathbb{N}}$ 122

3 Representation of Bounded Linear Operators

3.1 An integral representation of points of X 128
3.2 An integral representation for T 133
3.3 Compact operators revisited 136

Bibliography 141

Author Index 147

Subject Index 149

Notation Index 151

Preface

The main object of this book is to give a self-contained account of recent work ([34], [35], [36], [40], [41]) concerning the representation in series form of compact linear maps $T : X \to Y$, where X and Y are reflexive Banach spaces with strictly convex duals. When X and Y are Hilbert spaces such a result is classical and is due to the work by Hilbert, F. Riesz and E. Schmidt in the early years of the twentieth century; it is an apotheosis of early operator theory. Outside Hilbert spaces the problems encountered by the lack of orthogonality are so severe that it is hardly surprising that some restrictions on X and Y are needed to make any progress. By use of polar sets and projections (in general, nonlinear) as substitutes for orthogonal complements the following are constructed:

(i) a decreasing sequence (X_n) of linear subspaces of X with finite codimension and trivial intersection if T has a trivial kernel;

(ii) a sequence (x_n) of points in the unit sphere of X such that the norm λ_n of the restriction of T to X_n is attained at x_n, and with a semi-orthogonality property in a Banach space sense;

(iii) a recursively calculable sequence of scalars $(\xi_n(x))$ analogous to the sequence of Fourier coefficients of x that appears in the Hilbert space case.

Despite the obvious parallels between the recursive process used in the Hilbert space case and that adopted here, there is in fact a significant difference: the eigenvalues of T or $|T|$ (the positive square root of T^*T) that play such an important part when X and Y are Hilbert spaces no longer appear. The numbers λ_n that replace them have dual characteristics, for they are both

(a) eigenvalues of a **nonlinear** operator,

and (which emphasises their **linear** origins)

(b) norms of restrictions of T to the subspaces X_n of X with finite codimension.

This tends to suggest that when leaving the safe and comfortable framework of Hilbert spaces, the classical spectrum of the linear operator T may lose some of its overwhelming importance insofar as the representation of T is concerned. Of course, addiction to classical eigenvalues is firmly rooted in mathematicians' psyche, but nevertheless we think that such a point of view deserves serious consideration even though it may serve as grist to the mill of those who consider Banach spaces that are not Hilbert spaces to be strange and exotic objects.

The λ_n and x_n correspond to an "eigenvalue" and "eigenvector" respectively of a nonlinear operator equation involving a duality map that becomes the identity map in the Hilbert spaces case. In fact, when X and Y are Hilbert spaces, the λ_n are eigenvalues of $|T|$ and (x_n) is orthonormal. The most attractive form of the results presented here, when T has trivial kernel, is that under an additional assumption,

$$x = \sum_n \xi_n(x)x_n \quad \text{and} \quad Tx = \sum_n \lambda_n \xi_n(x)y_n$$

for all $x \in X$; here $y_n = Tx_n/\|Tx_n\|$. Thus the x_n form a basis of X and the representation of T is directly comparable with that known for the Hilbert space situation. The additional hypothesis is that, with $S_n x := \sum_{j=1}^{n-1} \xi_j(x)x_j \ (x \in X)$,

$$\sup_n \|S_n\| < \infty. \tag{H}$$

This condition is satisfied whenever X is a Hilbert space, no matter what Y is (within the class considered), and also in certain other cases. We do not know the extent of spaces and operators for which (H) holds, but if X is so unpleasant that it does not have the approximation property, there can be no compact linear map $T : X \to Y$ (whatever Y is, within the class considered) with trivial kernel for which (H) is satisfied, for otherwise X would have a basis, in contradiction to the lack of the approximation property.

Representations of $x \in X$ and Tx are also derived without hypothesis (H), but they are generally less elegant and involve nonlinear projections. However, additional progress can be made via the Gelfand numbers $c_n(T)$ of T: we recall that these are defined by

$$c_n(T) = \inf\left\{ \|TJ_M^X\| : \operatorname{codim} M < n \right\},$$

where J_M^X is the natural embedding from the closed linear subspace M of X into X. It turns out that if the $c_n(T)$ decay sufficiently quickly (2^{-2n} will do), then a Y-analogue of (H) holds, from which may be obtained a series representation of Tx, though not of x. We emphasise that, solely on account of this rapid decay and now standard theory, the map T is nuclear and hence may be expressed as a series: the point of the result presented here is that the series representation given involves coefficients recursively calculable by procedures broadly similar to those used in the Hilbert space case.

Chapter 1 provides the basic facts concerning the geometry of Banach spaces, orthogonality, bases, the approximation property and p-trigonometric functions that are needed later; the treatment is virtually self-contained as the proofs of most assertions are given in full. It also presents a brief account of s-numbers and entropy numbers in which, principally for reasons of space, detailed proofs are eschewed in favour of precise references. However, particular attention is paid to the Gelfand numbers $c_n(T)$ and the Gelfand widths $\tilde{c}_n(T)$ of T, defined by

$$\tilde{c}_n(T) = \inf_{L_n} \sup \left\{ \|Tx\|_Y : \|x\|_X \le 1, Tx \in L_n \right\},$$

where the infimum is taken over all closed linear subspaces L_n of Y with codimension at most $n - 1$. In general these two sets of quantities can be very different, but we show that if both X and Y are uniformly convex and uniformly smooth, and T has trivial kernel and range dense in Y, then $c_n(T) = \tilde{c}_n(T)$ for all $n \in \mathbb{N}$.

This machinery is deployed in Chapter 2 to give the results mentioned above concerning the representation in series form of compact operators. When (H) holds, not only do the x_n form a basis of X, but (ξ_n) is a basis of X^* and, for each $x \in X$, the sequence $(\xi_n(x))$ belongs to l_q for some q expressible in terms of the basis constant of (ξ_n). This may be thought of as a weak version of the Hausdorff–Young theorem about Fourier coefficients.

The assumption that the Banach spaces X and Y have strictly convex duals means that their norms are Gâteaux-differentiable (except at the origin). This enables duality maps $J_X : X \to X^*$ and $J_Y : Y \to Y^*$ to be defined and these have a prominent role in the theory. Sobolev spaces $\overset{0}{W}{}^k_p(\Omega)$, and Lebesgue spaces $L_p(\Omega)$ defined on an open subset Ω of $\mathbb{R}^n$, and sequence spaces l_p, have strictly convex duals for $1 < p < \infty$ and explicit duality maps can be determined in each case. For instance, for $\overset{0}{W}{}^k_p(\Omega)$, the duality map is the p-Laplacian when $k = 1$ and the p-biharmonic operator when $k = 2$. Moreover, if Ω is bounded, the embedding $\overset{0}{W}{}^k_p(\Omega) \hookrightarrow L_q(\Omega)$ is compact for a well-known range of values of k, p, q, and taking T to be this embedding, the abstract theory yields a rich vein of results. As a sample take $k = 1$, $1 < p < \infty$, $q = p$: the abstract results apply to establish the existence of an infinite sequence of "eigenfunctions" x_n and "eigenvalues" λ_n of the Dirichlet problem on Ω for the p-Laplacian:

$$-\Delta_p u := -\sum_{j=1}^{n} D_j \left(|D_j u|^{p-2} D_j u \right) = \gamma |u|^{p-2} u, \quad \text{on } \Omega, \quad u = 0 \text{ on } \partial\Omega.$$

This is satisfied in a weak sense relative to a decreasing family of subspaces X_n of $\overset{0}{W}{}^1_p(\Omega)$, and the "eigenfunctions" and "eigenvalues" are respectively the critical points and critical levels of an associated functional. The x_n satisfy a semi-orthogonality condition investigated by James in [52] and are referred to as j-eigenfunctions; the λ_n are called j-eigenvalues. Similar implications of the abstract results follow for other values of k and q. In the one-dimensional case of the p-Laplacian the actual eigenfunctions and eigenvalues are known and these coincide with those obtained by the Lusternik–Schnirelmann procedure. This is also shown to be so more generally for a Sturm–Liouville operator which is similar in form to the p-Laplacian. However, these eigenfunctions do not have the semi-orthogonality property of James and the eigenvalues are not equal to the j-eigenvalues. The eigenvalue problem for a compact Hardy-type operator T on $L_p(0, \infty)$ reduces to a Sturm–Liouville eigenvalue problem, and in this case, a Prüfer-type transformation is on hand to help prove that the eigenvalues coincide with the approximation numbers of T.

The final chapter is concerned with the problems encountered when seeking to extend the earlier results to non-compact maps. Integral representations are obtained for any $x \in X$ and Tx in terms of a right-continuous, non-decreasing family of projection operators $(P_\lambda)_{\lambda>0}$ (a resolution of the identity), and, in particular, if $\ker(T) = \{0\}$ and X is uniformly convex,

$$x = \int_{(0, \|T\|]} dP_\lambda x, \quad Tx = \int_{(0, \|T\|]} dT P_\lambda x.$$

When X is a Hilbert space, we derive an integral representation

$$\langle Tx, J_X Tx \rangle_Y = \int_{(0, \|T\|]} f(\lambda; x) d(P_\lambda x, x)_X$$

for each $x \in X$, where $\langle y, y^* \rangle_Y$ denotes the value of $y^* \in Y^*$ at $y \in Y$ and $f(\cdot; x) \in L_1((0, \|T\|])$. The difficulties faced in trying to find more specific information about $f(\cdot; x)$ and to establish such a result for a general X are underlined by the fact that even in the Hilbert space case, such an integral representation is available only for normal operators, and as Dunford and Schwarz point out in [31], the theory of normal operators is not an infallible guide to the theory for more general operators, even in finite-dimensional spaces. However, in line with our remarks above it seems possible to us that the problems in that situation are compounded by an insistence that the decomposition be expressed in spectral terms. Such a view appears to be in line with that of Davies in [24], who remarks that for non-self-adjoint problems "one must give up any hope that theorems about self-adjoint operators will provide useful signposts: they regularly lead in the wrong direction"; and that the significance of eigenvalues becomes a moot point. In our Banach space environment such comments have added force. Here we report on preliminary work in the hope of stimulating further attacks on this challenging problem.

This book has its roots in various papers written by the authors together with Desmond Harris; the incisive contributions of Jan Lang have also been of great importance. It is a pleasure to acknowledge here the many stimulating discussions we have had with both of them over the years.

Basic Notation

$\mathbb{R}$: real numbers.
$\mathbb{R}^n$: n-dimensional Euclidean space.
$\mathbb{C}$: complex numbers.
$\mathbb{Z}$: integers.
$\mathbb{N}$: natural numbers.
$\mathbb{N}_0 = \mathbb{N} \cup \{0\}$.
$\mathbb{Q}$: rational numbers.
$\delta_{i,j}$: Kronecker delta.
sgn: signum function: $\operatorname{sgn} 0 = 0$, $\operatorname{sgn} z = z/|z|$ if $z \neq 0$.
χ_E: characteristic function of E.
$\ker(T)$: kernel of a map T.
$\upharpoonright_E$: restriction to E.
$A \subset B$: A contained in B, or possibly equal to B.
Ω: open subset of $\mathbb{R}^n$.
$\partial\Omega$: boundary of Ω.
$\overline{\Omega}$: closure of Ω.
$\|\cdot\|_X$ or $\|\cdot|X\|$: norm on X.
X, Y: Banach spaces with duals X^*, Y^*.
B_X: closed unit ball in X.
S_X: unit sphere in X.
$\operatorname{sp} S$: linear span of $S \subset X$.
$\overline{\operatorname{sp} S}$: closed linear span of S.
$B(X,Y)$ ($B(X)$ when $Y = X$): space of bounded linear operators from X to Y.
$\langle x, x^* \rangle_X$: value of $x^* \in X^*$ at $x \in X$.
$\rightharpoonup$: weak convergence.
$\rightharpoonup^*$: weak* convergence.
$X \hookrightarrow Y$: X is continuously embedded in Y.
$L_p(\Omega)$, $1 < p < \infty$: Lebesgue space of functions f with $|f|^p$ integrable on Ω.
$\|\cdot\|_p$ or $\|\cdot\|_{p,I}$: norm on $L_p(I)$.
l_p, $1 < p < \infty$: space of sequences $(x_n)_{n \in \mathbb{N}}$ such that $\sum_{n=1}^{\infty} |x_n|^p < \infty$.
$W_p^r(\Omega)$: Sobolev spaces.
$C_0^\infty(\Omega)$: infinitely differentiable functions with compact supports in Ω.
$\overset{0}{W}{}_p^r(\Omega)$: closure of $C_0^\infty(\Omega)$ in $W_p^r(\Omega)$.

Chapter 1

Preliminaries

In this chapter we collect information about the geometry of Banach spaces, Schauder bases and p-trigonometric functions that will be useful later on. For the convenience of the reader, proofs of most assertions are given, precise references being provided for the remainder.

As a matter of notation throughout the book, the norm on a normed linear space space X will usually be written as $\|\cdot \mid X\|$ or $\|\cdot\|_X$, depending on the size of the expression X; if no ambiguity is likely we shall simply write $\|\cdot\|$, and often $\|\cdot\|_p$ or $\|\cdot\|_{p,I}$ will be used to represent the norm on $L_p(I)$; by $\dim X$ we mean the dimension of X; and B_X (resp. S_X) will stand for the closed unit ball (resp. the unit sphere) in X. Given any set $S \subset X$, the linear span of S, $\operatorname{sp} S$, is the smallest linear subspace of X containing S; the closed linear span of S, $\overline{\operatorname{sp}} S$, has the obvious analogous definition. The space of all bounded linear maps from a normed linear space X to another such space Y is denoted by $B(X, Y)$, or $B(X)$ if $X = Y$, and we write $B(X, \mathbb{K}) = X^*$ (the dual of X) when $\mathbb{K}$ ($= \mathbb{R}$ or $\mathbb{C}$) is the space of scalars associated with X. If X and Y are Banach spaces, $B(X, Y)$ is a Banach space with respect to the norm $\|\cdot\|$ defined by $\|T\| = \sup\{\|Tx\|_Y : \|x\|_X \leq 1\}$. The value of $x^* \in X^*$ at $x \in X$ is denoted by $\langle x, x^*\rangle_X$ or $\langle x, x^*\rangle$. Weak convergence in X is represented by a half arrow $\rightharpoonup$, weak* convergence in X^* by $\overset{*}{\rightharpoonup}$. Unless special mention is made to the contrary, all statements and proofs hold for both the real and the complex scalar fields. A linear operator $T : X \to Y$, acting between Banach spaces X, Y, is defined to be compact if, for any bounded set B in X, $T(B)$ is precompact, i.e., its closure $\overline{T(B)}$ is compact in Y. Equivalently, given a bounded sequence (x_n) in X, the sequence (Tx_n) contains a subsequence which converges in Y. A compact linear operator is necessarily bounded, and the family of compact linear maps from X to Y is closed in $B(X, Y)$.

1.1   The geometry of Banach spaces

Definition 1.1.1. A Banach space X is said to be **strictly convex** if whenever $x, y \in X$ are such that $x \neq y$ and $\|x\| = \|y\| = 1$, and $\lambda \in (0, 1)$, then $\|\lambda x + (1 - \lambda) y\| < 1$.

This means that no sphere in X contains a line segment. Use of the conditions for equality in Minkowski's inequality shows that the spaces l_p and L_p are strictly convex if $1 < p < \infty$ and do not have this property if p is 1 or ∞. Evidently every linear subspace of a strictly convex space is strictly convex with respect to the inherited norm. The next result gives a simple condition equivalent to strict convexity.

Proposition 1.1.2. *A Banach space X is strictly convex if and only if it is the case that whenever $x, y \in X$ are such that $\|x + y\| = \|x\| + \|y\|$ then either $y = 0$ or $x = \lambda y$ for some $\lambda \geq 0$.*

Proof. Suppose that X is strictly convex and $x, y \in X \backslash \{0\}$ are such that $x \neq y$ and $\|x + y\| = \|x\| + \|y\|$. Thus $\|x\| \neq \|y\|$, for otherwise $\left\| \frac{x+y}{2} \right\| = \|x\|$, contradicting the strict convexity of X. However, if $\|y\| < \|x\|$, then with $\lambda = \|y\| / \|x\|$ we have

$$1 \geq \left\| \frac{x}{\|x\|} + \lambda \left(\frac{y}{\|y\|} - \frac{x}{\|x\|} \right) \right\| \geq \frac{\|x + y\| - \lambda \|x\|}{\|x\|} = \frac{\|x\| + \|y\| - \lambda \|x\|}{\|x\|} = 1,$$

which means that $x = y/\lambda$. The converse is obvious. $\square$

At a deeper level we have

Proposition 1.1.3. *Let X be a Banach space. Then X is strictly convex if and only if given any $x^* \in X^* \backslash \{0\}$, there exists at most one $x \in X$ such that $\|x\| = 1$ and $\langle x, x^* \rangle = \|x^*\|$; such an x exists if X is reflexive.*

Proof. Let X be strictly convex and $x^* \in X^* \backslash \{0\}$. Suppose there are two distinct such points x, say x_1 and x_2. Then if $0 < \lambda < 1$,

$$\|x^*\| = \lambda \langle x_1, x^* \rangle + (1 - \lambda) \langle x_2, x^* \rangle = \langle \lambda x_1 + (1 - \lambda) x_2, x^* \rangle$$
$$\leq \|x^*\| \, \|\lambda x_1 + (1 - \lambda) x_2\| < \|x^*\|,$$

which is absurd. Conversely, suppose that $\|x + \lambda(y - x)\| = 1$ for some $x, y \in X$ with $\|x\| = \|y\| = 1$ and some $\lambda \in (0, 1)$. By the Hahn–Banach theorem, there exists $x^* \in X^*$ such that $\langle x + \lambda(y - x), x^* \rangle = 1$ and $\|x^*\| = 1$. Then $(1 - \lambda) \langle x, x^* \rangle + \lambda \langle y, x^* \rangle = 1$, and since $|\langle x, x^* \rangle|, |\langle y, x^* \rangle| \leq 1$ we must have $\langle x, x^* \rangle = \langle y, x^* \rangle = 1$. By hypothesis, this implies that $x = y$, and so X is strictly convex.

To prove that such an x exists when X is reflexive, let (x_k) be a sequence in X such that $\|x_k\| = 1$ for all $k \in \mathbb{N}$ and $\|x^*\| = \lim_{k \to \infty} \langle x_k, x^* \rangle$. Then there is a weakly convergent subsequence of (x_k), still denoted by (x_k) for simplicity, with weak limit x, say. Hence $\|x\| \leq 1$ and $\langle x, x^* \rangle = \lim_{k \to \infty} \langle x_k, x^* \rangle = \|x^*\|$. $\square$

The existence of a unique element of minimal norm in a convex, closed, non-empty subset of a reflexive, strictly convex space follows quickly.

Proposition 1.1.4. *Let K be a closed, convex, non-empty subset of a strictly convex Banach space X. Then there is at most one element $x \in K$ such that*

$$\|x\| = \inf\{\|y\| : y \in K\}.$$

If in addition X is reflexive, such an x exists.

Proof. Suppose there exist $x, y \in K$ with $\|x\| = \|y\| = \inf\{\|z\| : z \in K\}$, $x \neq y$. Let $0 < \lambda < 1$: then $\lambda x + (1 - \lambda)y \in K$, $\|\lambda x + (1 - \lambda)y\| < \|x\|$ and we have a contradiction.

Now let X be reflexive and assume that (x_k) is a sequence in K such that $\lim_{k \to \infty} \|x_k\| = l := \inf\{\|y\| : y \in K\}$. By reflexivity, this sequence has a subsequence, still denoted by (x_k) for convenience, such that $x_k \rightharpoonup x$ for some $x \in X$; in fact, $x \in K$ since K is convex and closed, and hence weakly closed. Moreover, $\|x\| \leq \lim_{k \to \infty} \|x_k\| = l$. □

Next we introduce an important class of strictly convex spaces.

Definition 1.1.5. The **modulus of convexity** $\delta_X : (0, 2] \to [0, 1]$ of a Banach space X with $\dim X \geq 2$ is defined by

$$\delta_X(\varepsilon) = \inf \left\{ 1 - \frac{1}{2} \|x + y\| : x, y \in B_X, \|x - y\| \geq \varepsilon \right\}.$$

The space X is called **uniformly convex** if $\delta_X(\varepsilon) > 0$ for all $\varepsilon \in (0, 2]$.

Remark 1.1.6.

(i) Clearly

$$\delta_X(\varepsilon) = \inf \left\{ \delta_Y(\varepsilon) : Y \text{ is a 2-dimensional subspace of } X \right\}.$$

Moreover, δ_X is increasing on $[0, 2]$, continuous on $[0, 2)$ (not necessarily at 2) and $\delta_X(0) = 0$; in addition, $\delta_X(2) = 1$ if and only if X is strictly convex.

(ii) The modulus of convexity of X is also given by the formula

$$\delta_X(\varepsilon) = \inf \left\{ 1 - \frac{1}{2} \|x + y\| : x, y \in X, \|x\| = \|y\| = 1, \|x - y\| = \varepsilon \right\}.$$

To establish this, note that if $x, y \in X$ are such that $\|x\| \leq 1, \|y\| \leq 1$ and $\|x - y\| \geq \varepsilon$, then there are points x_1, y_1 on the line segment joining x to y such that $(x_1 + y_1)/2 = (x + y)/2$ and $\|x_1 - y_1\| = \varepsilon$. In the computation of $\delta_X(\varepsilon)$ it is therefore enough to consider only those points $x, y \in B_X$ such that $\|x - y\| = \varepsilon$. All that is left is to show that

$$K_B := \sup \left\{ \|x + y\| : x, y \in B_X, \|x - y\| = \varepsilon \right\}$$

coincides with

$$K_S := \sup \{ \|x + y\| : x, y \in S_X, \|x - y\| = \varepsilon \} .$$

In view of (i) it is sufficient to deal with the case when X is two-dimensional. Then the suprema are attained; suppose that K_B is attained at $x_0, y_0 \in B_X$. We wish to prove that $x_0, y_0 \in S_X$. Assume that $\|y_0\| < 1$, put $A = \{z \in B_X : \|z - x_0\| = \varepsilon\}$ and let $x^* \in X^*$ be such that $\|x^*\| = 1$ and $\langle x_0 + y_0, x^* \rangle_X = \|x_0 + y_0\|$. Then for all $z \in A$,

$$\operatorname{re} \langle x_0 + z, x^* \rangle_X \le \|x_0 + z\| \le \|x_0 + y_0\| = \langle x_0 + y_0, x^* \rangle_X .$$

Hence $\operatorname{re} x^*$ attains its supremum on A at y_0, so that by translation,

$$\operatorname{re} \langle y_0 - x_0, x^* \rangle_X = \|y_0 - x_0\| = \varepsilon.$$

(Note that it is easy to see that for all $f \in X^*$,

$$\sup_{x \in B_X} |\langle x, f \rangle_X| = \sup_{x \in B_X} |\operatorname{re} \langle x, f \rangle_X| .)$$

Thus

$$\|x_0\| \le \frac{1}{2} \left(\|x_0 + y_0\| + \|y_0 - x_0\| \right)$$
$$= \frac{1}{2} \left(\operatorname{re} \langle x_0 + y_0, x^* \rangle_X + \operatorname{re} \langle y_0 - x_0, x^* \rangle_X \right)$$
$$= \operatorname{re} \langle y_0, x^* \rangle_X < 1.$$

However, with $\eta := \frac{1}{2} \min \{ 1 - \|x_0\|, 1 - \|y_0\| \} > 0$ we have

$$x_1 := x_0 + \eta(x_0 + y_0), y_1 := y_0 + \eta(x_0 + y_0) \in B_X, \|x_1 - y_1\| = \varepsilon$$

and

$$\|x_1 + y_1\| = (1 + 2\eta) \|x_0 + y_0\| > \|x_0 + y_0\| ,$$

which contradicts the maximality of $\|x_0 + y_0\|$. It follows that $\|y_0\| = 1$; a similar argument gives $\|x_0\| = 1$.

(iii) Each of the following two conditions is equivalent to the hypothesis of uniform convexity of X:

  (a) If (x_n), (y_n) are sequences in X such that (x_n) is bounded and

$$\lim_{n \to \infty} \left(2 \|x_n\|^2 + 2 \|y_n\|^2 - \|x_n + y_n\|^2 \right) = 0,$$

   then $\lim_{n \to \infty} \|x_n - y_n\| = 0$.
  (b) If for all $n \in \mathbb{N}$ there are points $x_n, y_n \in B_X$ such that $\lim_{n \to \infty} \|x_n + y_n\| = 2$, then $\lim_{n \to \infty} \|x_n - y_n\| = 0$.

We prove only the implication (b) $\implies$ (a) as the remaining parts are fairly routine. Suppose that $(x_n), (y_n)$ are as in (a). Since

$$2 \left\| x_n \right\|^2 + 2 \left\| y_n \right\|^2 - \left\| x_n + y_n \right\|^2 \geq 2 \left\| x_n \right\|^2 + 2 \left\| y_n \right\|^2 - (\left\| x_n \right\| + \left\| y_n \right\|)^2$$
$$= (\left\| x_n \right\| - \left\| y_n \right\|)^2 \geq 0,$$

we see that $\lim_{n \to \infty} (\left\| x_n \right\| - \left\| y_n \right\|) = 0$: the sequence (y_n) must be bounded. Passage to a subsequence, if necessary, enables us to assume that $\lim_{n \to \infty} \left\| x_n \right\| = \lim_{n \to \infty} \left\| y_n \right\| = a$, say. There is nothing further to prove if $a = 0$; suppose that $a > 0$. Then $\left\| x_n + y_n \right\| \to 2a$; and $x / \left\| x_n \right\|, y / \left\| y_n \right\| \in B_X$ for all large enough n, while

$$\left\| \frac{x_n}{\left\| x_n \right\|} + \frac{y_n}{\left\| y_n \right\|} \right\| \to 2.$$

If (b) holds, then

$$\left\| \frac{x_n}{\left\| x_n \right\|} - \frac{y_n}{\left\| y_n \right\|} \right\| \to 0,$$

so that $\left\| x_n - y_n \right\| \to 0$ and (a) is satisfied.

(iv) Every closed linear subspace of a uniformly convex space is uniformly convex when given the inherited norm. While every uniformly convex space is strictly convex, the converse is false, in general (see [25]). However, if X is strictly convex and $\dim X < \infty$, then X is uniformly convex, for the fact that $1 - \frac{1}{2} \left\| x + y \right\| > 0$ whenever $\left\| x \right\| = \left\| y \right\| = 1$ and $\left\| x - y \right\| = \varepsilon > 0$ implies that $\delta_X(\varepsilon) > 0$ since closed bounded sets in X are compact.

(v) Every Hilbert space H is uniformly convex as, with the use of the parallelogram law, we have

$$\delta_H(\varepsilon) = \inf \left\{ 1 - \left(\frac{\left\| x \right\|^2}{2} + \frac{\left\| y \right\|^2}{2} - \left\| \frac{x - y}{2} \right\|^2 \right)^{1/2} : \left\| x \right\| = \left\| y \right\| = 1, \left\| x - y \right\| = \varepsilon \right\}$$
$$= 1 - (1 - \varepsilon^2/4)^{1/2} > 0 \text{ for all } \varepsilon \in (0, 2].$$

Note that since

$$(1 - x^q)^{1/q} \leq 1 - x^q/q \quad (0 \leq x \leq 1, 1 < q < \infty),$$

we have

$$\delta_H(\varepsilon) \geq \varepsilon^2/8.$$

In the opposite direction, for every Banach space X of dimension at least 2, it is known that

$$\delta_X(\varepsilon) \leq \delta_H(\varepsilon) = 1 - (1 - \varepsilon^2/4)^{1/2} \leq C\varepsilon^2;$$

see [62], Vol. II, p. 63. In this sense, Hilbert spaces are the 'most' uniformly convex spaces.

To extend our repertoire of uniformly convex spaces we consider the Lebesgue spaces $L_p = L_p(\Omega, \mu)$, where $p \in (1, \infty)$, (Ω, μ) is a measure space and the norm is given by

$$\|x\|_p := \left(\int_\Omega |x|^p \, d\mu \right)^{1/p}.$$

The modulus of convexity of L_p is denoted by δ_p. That such spaces are uniformly convex was established by Clarkson [23]. In this connection a preliminary lemma will be useful (see [5] and [39], p. 6).

Lemma 1.1.7. *Let $1 < p < \infty$ and define $\lambda_p : [0, \infty) \to [0, \infty)$ by*

$$\lambda_p(t) = (1 + t)^{p-1} + |1 - t|^{p-1} \operatorname{sgn}(1 - t).$$

Then for all $x, y \in \mathbb{R}$,

$$|x + y|^p + |x - y|^p = \sup\{\lambda_p(t) |x|^p + \lambda_p(1/t) |y|^p : 0 < t < \infty\}$$

if $1 < p \le 2$; if $2 \le p < \infty$, the same holds with sup replaced by inf.

Proof. Suppose that $1 < p \le 2$. By homogeneity and symmetry it is enough to deal with the case $0 < y \le x = 1$. Put

$$f(t) = \lambda_p(t) + \lambda_p(1/t)y^p, \quad 0 < t < \infty;$$

thus $f(y) = (1 + y)^p + (1 - y)^p$. If $t \ne 1$,

$$f'(t) = (p - 1)\{1 - (y/t)^p\} \left\{ (1 + t)^{p-2} - |1 - t|^{p-2} \right\},$$

so that $f'(t) \ge 0$ if $0 < t < y$, $f'(t) \le 0$ if $y < t < \infty$ ($t \ne 1$). Hence the maximum value of f on $(0, \infty)$ occurs when $t = y$: the required inequality follows. The proof when $2 \le p < \infty$ is similar. $\qquad\square$

Theorem 1.1.8. *Let $f, g \in L_p$. Then if $1 < p \le 2$,*

$$\|f + g\|_p^p + \|f - g\|_p^p \ge \left(\|f\|_p + \|g\|_p \right)^p + \left| \|f\|_p - \|g\|_p \right|^p,$$

while if $2 \le p < \infty$ the inequality is reversed.

Proof. Suppose that $1 < p \le 2$. Then by Lemma 1.1.7, all integrals being over Ω,

$$\|f + g\|_p^p + \|f - g\|_p^p = \int \{|f + g|^p + |f - g|^p\} \, d\mu$$

$$= \int \sup_{0 < t < \infty} \{\lambda_p(t) |f|^p + \lambda_p(1/t) |g|^p\} \, d\mu$$

$$\ge \sup_{0 < t < \infty} \int \{\lambda_p(t) |f|^p + \lambda_p(1/t) |g|^p\} \, d\mu$$

$$= \sup_{0 < t < \infty} \left\{ \lambda_p(t) \|f\|_p^p + \lambda_p(1/t) \|g\|_p^p \right\}$$

$$= \left(\|f\|_p + \|g\|_p \right)^p + \left| \|f\|_p - \|g\|_p \right|^p,$$

the final step following from another application of Lemma 1.1.7. The case $2 \le p < \infty$ is similar. □

Corollary 1.1.9. *When* $2 \le p < \infty$ *the space* L_p *is uniformly convex and its modulus of convexity satisfies* $\delta_p(\varepsilon) \ge (\varepsilon/2)^p/p$ $(\varepsilon \in (0,2))$.

Proof. Theorem 1.1.8 together with Remark 1.1.6 (ii) immediately give

$$\delta_p(\varepsilon) \ge 1 - (1 - (\varepsilon/2)^p)^{1/p} \ge 1 - (1 - (\varepsilon/2)^p/p) = (\varepsilon/2)^p/p.$$

□

Further work is needed to obtain a result of comparable sharpness when $1 < p \le 2$: we use the following result of Ball and Pisier (see [5]).

Theorem 1.1.10. *Suppose that* $1 < p \le 2$ *and let* $f, g \in L_p$. *Then*

$$\|f + g\|_p^2 + \|f - g\|_p^2 \ge 2 \|f\|_p^2 + 2(p - 1) \|g\|_p^2.$$

If $2 \le p < \infty$ *the inequality is reversed.*

Proof. Suppose that $1 < p \le 2$. We claim that

$$|a + b|^p + |a - b|^p \ge 2\{a^2 + (p - 1)b^2\}^{p/2} \ (a, b \in \mathbb{R}), \tag{1.1.1}$$

to prove which it is enough to show that

$$h(t) := (1 + t)^p + (1 - t)^p - 2\{1 + (p - 1)t^2\}^{p/2} \ge 0, \ 0 \le t \le 1.$$

Use of the binomial expansion gives

$$h(t) = 2p(p - 1) \sum_{n=2}^{\infty} \frac{(3 - p) \cdots (2n - 1 - p)}{(2n)!} t^{2n} + I,$$

where I is equal to

$$2 \sum_{m=1}^{\infty} \frac{p(2 - p) \cdots (4m - 2 - p)}{(2m + 1)! 2^{2m+1}} \{2(2m + 1) - (4m - p)(p - 1)t^2\} (p - 1)^{2m} t^{4m}.$$

Since all the terms involved in the series for I are non-negative when $0 \le t < 1$, (1.1.1) follows. By Hölder's inequality,

$$\|f + g\|_p^p + \|f - g\|_p^p \le 2^{(2-p)/2} \left(\|f + g\|_p^2 + \|f - g\|_p^2 \right)^{p/2},$$

from which we have, by Theorem 1.1.8,

$$\left(\|f + g\|_p^2 + \|f - g\|_p^2 \right)^{p/2} \ge 2^{-(2-p)/2} \left\{ \left(\|f\|_p + \|g\|_p \right)^p + \left| \|f\|_p - \|g\|_p \right|^p \right\}$$

$$\ge 2^{p/2} \left\{ \|f\|_p^2 + (p - 1) \|g\|_p^2 \right\}^{p/2},$$

the final inequality following from (1.1.1). This gives the theorem when $1 < p \le 2$; similar arguments work when $2 \le p < \infty$. □

Corollary 1.1.11. *When* $1 < p \leq 2$, *the space* L_p *is uniformly convex and its modulus of convexity satisfies* $\delta_p(\varepsilon) \geq (p-1)\varepsilon^2/8$ ($\varepsilon \in (0,2)$).

Proof. From the last theorem, with $u = f + g$ and $v = f - g$, we have

$$\|u\|_p^2 + \|v\|_p^2 \geq 2 \left\| \frac{u+v}{2} \right\|_p^2 + 2(p-1) \left\| \frac{u-v}{2} \right\|_p^2$$

for all $u, v \in L_p$. Hence

$$\delta_p(\varepsilon) \geq 1 - \left\{ 1 - (p-1)(\varepsilon/2)^2 \right\}^{1/2} \geq (p-1)\varepsilon^2/8. \qquad \square$$

Remark 1.1.12. As a special case of Corollaries 1.1.9 and 1.1.11, it follows that if $1 < p < \infty$, then the sequence space l_p and its n-dimensional analogue l_p^n ($n \in \mathbb{N}$) are uniformly convex. Since $\prod_{k=1}^n L_p$, endowed with the norm

$$\|(f_1, \ldots, f_n)\| := \left(\sum_{k=1}^n \|f_k\|_p^p \right)^{1/p},$$

may be thought of as a space of l_p^n form, it can be seen that $\prod_{k=1}^n L_p$ is also uniformly convex, with the same modulus of convexity as L_p.

We next give some useful properties of uniformly convex spaces.

Proposition 1.1.13. *Let* (x_k) *be a sequence in a uniformly convex Banach space* X *that converges weakly to* $x \in X$, *with* $\|x_k\| \to \|x\|$. *Then* $\|x_k - x\| \to 0$.

Proof. As the result is trivial if $x = 0$ we assume that $x \neq 0$; we may plainly also assume that for all $k \in \mathbb{N}$, $x_k \neq 0$. Put $\xi_k = 1 - \|x\| / \|x_k\|$; $\xi_k \to 0$ as $k \to \infty$. Set $y_k = x_k / \|x_k\|$, $y = x / \|x\|$ and note that $y_k = (1 - \xi_k) x_k / \|x\|$ converges weakly to y. Moreover, $\|y_k\| = \|y\| = 1$. By the Hahn–Banach theorem, there exists $y^* \in X^*$ such that $\langle y, y^* \rangle_X = 1 = \|y^*\|$. Hence

$$2 \geq \|y_k + y\| \geq |\langle y_k + y, y^* \rangle_X| \to 2 \langle y, y^* \rangle_X = 2,$$

so that $\lim_{k \to \infty} \|y_k + y\| = 2$. Since X is uniformly convex, $\lim_{k \to \infty} \|y_k - y\| = 0$, and as $x_k - x = \|x_k\| y_k - \|x\| y$, it follows easily that $\|x_k - x\| \to 0$. $\qquad \square$

Theorem 1.1.14. *Every uniformly convex Banach space is reflexive.*

Proof. This classical result is due to D.P. Milman [67]; here we give the short proof contained in [62], Vol. II. Suppose that X is uniformly convex, let $J : X \to X^{**}$ be the canonical map, let B, B^{**} be the closed unit balls in X, X^{**} respectively and denote by σ the weak topology $\sigma(X^{**}, X^*)$. Let $x^{**} \in X^{**}$, $\|x^{**}\| = 1$. Then since $J(B)$ is σ-dense in B^{**} (see [11], Chap. IV, § 5), there is a generalised sequence

$\{x_\alpha\}_{\alpha \in A}$ $(x_\alpha \in B)$ such that $Jx_\alpha \overset{\sigma}{\to} x^{**}$. Since $Jx_\alpha + Jx_\beta \overset{\sigma}{\to} 2x^{**}$, the weak lower-semicontinuity of the norm shows that $\lim_{\alpha,\beta} \|Jx_\alpha + Jx_\beta\| = 2$, and hence

$$\lim_{\alpha,\beta} \|x_\alpha + x_\beta\| = 2. \tag{1.1.2}$$

As X is uniformly convex, it follows from (1.1.2) that $\lim_{\alpha,\beta} \|x_\alpha - x_\beta\| = 0$, so that for some $x \in B$, $\lim_\alpha \|x_\alpha - x\| = 0$. Thus $\lim_\alpha \|Jx_\alpha - Jx\| = 0$ and hence $x^{**} = Jx$: X must be reflexive. $\qquad\square$

Remark 1.1.15. A simple proof of the last theorem can be given via a celebrated theorem of James [53]: a Banach space X is reflexive if and only if every $f \in X^*$ attains its norm. Unfortunately the proof of this result is too long for inclusion here. However, assuming it and supposing that X is a uniformly convex Banach space, we proceed as follows. Let $f \in X^*$ have unit norm and let (x_n) be a sequence in the unit sphere of X such that $\lim_{n\to\infty} \langle x_n, f \rangle = 1$. Given $\varepsilon > 0$, there exists $N \in \mathbb{N}$ such that

$$2 - \varepsilon \le |\langle x_n + x_m, f \rangle| \le \|x_n + x_m\| \le 2 \; (m, n \ge N).$$

Since X is uniformly convex, (x_n) is a Cauchy sequence and so converges to a point x in the unit sphere of X; plainly $\langle x, f \rangle = 1$, so that f attains its norm. The James theorem now ensures that X is reflexive.

As an application of uniform convexity we consider projections. Let K be a closed, convex, non-empty subset of a uniformly convex Banach space X. Given any $x \in X$, the set $K - \{x\} := \{y - x : y \in K\}$ is also closed, convex and non-empty, and so by Proposition 1.1.4, it has a unique element $w(x)$ of minimal norm. Since $w(x) = z(x) - x$ for some unique $z(x) \in K$, we have

$$\|z(x) - x\| = \inf\{\|y - x\| : y \in K\} = d(x, K).$$

Hence there is a unique point $P_K x := z(x)$ that is the nearest point of K to x. The map $P_K : X \to K$ thus defined is called the **projection of X onto K** as clearly $P_K x = x$ if and only if $x \in K$. In general, P_K is nonlinear. If X is a Hilbert space, it is easy to show that P_K is Lipschitz-continuous with constant 1:

$$\|P_K x - P_K y\| \le \|x - y\| \text{ for all } x, y \in X.$$

Such a result is not available in general, but we do have

Proposition 1.1.16. *Let K be a closed, convex, non-empty subset of a uniformly convex Banach space X. Then the projection map P_K is continuous.*

Proof. Suppose the result is false. Then there exist $x \in X$, a sequence (x_n) in X and $\varepsilon > 0$ such that $\lim_{n\to\infty} x_n = x$ and $\|P_K x_n - P_K x\| \ge \varepsilon$ for all $n \in \mathbb{N}$. Since the distance $d(x, K)$ from x to K satisfies $|d(x, K) - d(x_n, K)| \le \|x_n - x\|$, it follows that

$$\big|\|x_n - P_K x_n\| - \|x - P_K x\|\big| \le \|x_n - x\| \to 0 \text{ as } n \to \infty.$$

As $(P_K x_n)$ is bounded, it has a weakly convergent subsequence, still denoted by $(P_K x_n)$ for convenience, with weak limit z, say. Since K is closed and convex, it is weakly closed and so $z \in K$. Moreover, $x_n - P_K x_n \rightharpoonup x - z$ and

$$\|x - z\| \le \liminf_{n \to \infty} \|x_n - P_K x_n\| = \|x - P_K x\|,$$

which implies that $z = P_K x$. Hence

$$x_n - P_K x_n \rightharpoonup x - P_K x \text{ and } \|x_n - P_K x_n\| \to \|x - P_K x\|.$$

By Proposition 1.1.13, the uniform convexity of X now implies that $x_n - P_K x_n \to x - P_K x$, so that $P_K x_n \to P_K x$: contradiction. $\square$

Further insight into strict and uniform convexity is provided by differentiability considerations. Let U be an open subset of a Banach space X and $f : U \to \mathbb{R}$. We say that f is **Gâteaux-differentiable** at $x_0 \in U$ if there exists $x^* \in X^*$ such that

$$\langle h, x^* \rangle = \lim_{t \to 0} \frac{f(x_0 + th) - f(x_0)}{t} \text{ for all } h \in X.$$

The limit above is called the **derivative of f in the direction h**; the functional x^* is denoted by $\operatorname{grad} f(x_0)$ and will be referred to as the **gradient** or **Gâteaux derivative** of f at x_0. If this limit is uniform with respect to $h \in S_X$, we say that f is **Fréchet-differentiable** at x_0 (and then refer to x^* as the **Fréchet derivative** of f at x_0, written $f'(x_0)$): equivalently,

$$\frac{f(x_0 + h) - f(x_0) - \langle h, x^* \rangle}{\|h\|} \to 0 \text{ as } \|h\| \to 0.$$

Clearly Fréchet-differentiability implies Gâteaux-differentiability; in the reverse direction, it can be shown that if $\operatorname{grad} f(x)$ exists throughout some neighbourhood of x_0 and is continuous at x_0, then f is Fréchet-differentiable at x_0 and $\operatorname{grad} f(x_0) = f'(x_0)$.

The case when $f(x) = \|x\|$ $(x \in X)$ is of particular interest. The norm $\|\cdot\|$ is said to be Fréchet- (resp. Gâteaux-) differentiable if it is Fréchet- (resp. Gâteaux-) differentiable at every point of $X \backslash \{0\}$: the point 0 is excluded because no norm is differentiable at 0. In view of the homogeneity property of a norm, to verify that $\|\cdot\|$ is (Fréchet- or Gâteaux-) differentiable it is plainly enough to establish differentiability at all points of S_X. Note that if $x, h \in X$ and $0 < s < t$, then the convexity of $y \longmapsto \|x + y\|$ implies that

$$\|x + sh\| - \|x\| = \left\| x + \frac{s}{t} th + \left(1 - \frac{s}{t}\right) \cdot 0 \right\| - \|x\| \le \frac{s}{t} (\|x + th\| - \|x\|),$$

which shows that $t \mapsto t^{-1}(\|x + th\| - \|x\|)$ is monotone increasing in $t > 0$. Moreover, $t^{-1}(\|x + th\| - \|x\|)$ is bounded below, for use of the triangle inequality shows that

$$t^{-1}(\|x + th\| - \|x\|) \ge -\|h\|.$$

Hence the one-sided limits

$$F_+(h) := \lim_{t \to 0+} t^{-1}(\|x + th\| - \|x\|) \text{ and } F_-(h) := \lim_{t \to 0-} t^{-1}(\|x + th\| - \|x\|)$$

exist for all $x, h \in X$. Since

$$\|x + th\| - \|x\| \geq \|x\| - \|x - th\|,$$

we see that $F_-(h) \leq F_+(h)$.

Lemma 1.1.17. *The norm $\|\cdot\|$ of a Banach space X is Fréchet-differentiable if and only if for all $x \in X\backslash\{0\}$,*

$$\lim_{t \to 0} \frac{\|x + th\| + \|x - th\| - 2\|x\|}{t} = 0$$

uniformly for $h \in S_X$. A similar result characterises Gâteaux-differentiability (with no uniformity requirement).

If the norm on $X \backslash \{0\}$ is Gâteaux-differentiable and $F(x) = \text{grad} \|x\|$, then $\|F(x)\| = 1$ and $\langle x, F(x) \rangle = \|x\|$ for all $x \in X$.

Proof. Let F_+, F_- be as above. If $\|\cdot\|$ is Fréchet-differentiable, then $F_-(h) = F_+(h)$ for all $h \in S_X$ and since, for all $t > 0$,

$$\frac{\|x + th\| - \|x\|}{t} - \frac{\|x - th\| - \|x\|}{-t} = \frac{\|x + th\| + \|x - th\| - 2\|x\|}{t},$$

the claim follows. Conversely, if the condition holds, then $F_-(h) = F_+(h)$ for all $h \in S_X$ (hence for all $h \in X$); and $F := F_+ = F_-$ is a linear functional since F_+ is subadditive, F_- is superadditive and both are positively homogeneous. To show that $F \in X^*$, fix $x \in X\backslash\{0\}$ and note that

$$\left| \frac{\|x + th\| - \|x\|}{t} \right| \leq 1$$

if $h \in S_X$ and $t \neq 0$. Hence $F(h) \leq 1$ for all $h \in S_X$. The argument for Gâteaux-differentiability is similar. The rest of the lemma is easily proved. $\square$

Remark 1.1.18. Given a convex subset C of a Banach space X, and a point $x \in C$, a functional $f \in X^*$ is called a **supporting functional** of C at x if $\|f\| = 1$ and $\langle x, f \rangle_X = \sup\{|\langle y, f \rangle|_X : y \in C\}$. From the arguments above we see that if the norm on X is Gâteaux-differentiable, so that $\langle x, \text{grad} \|x\| \rangle_X = \|x\|$ if $x \neq 0$, then $\text{grad} \|x\|$ is a supporting functional of B_X at every point x of $B_X\backslash\{0\}$.

Lemma 1.1.19. *Let X be a Banach space and let $x \in S_X$.*

(i) *The following three statements are equivalent:*
    (a) *The norm $\|\cdot\|$ is Fréchet-differentiable at x.*

(b) $\lim_{n\to\infty} \|f_n - g_n\| = 0$ whenever $f_n, g_n \in S_{X^*}$ satisfy $\lim_{n\to\infty} \langle x, f_n \rangle = \lim_{n\to\infty} \langle x, g_n \rangle = 1$.

(c) A sequence (f_n) in S_{X^*} is convergent whenever $\lim_{n\to\infty} \langle x, f_n \rangle = 1$.

(ii) The following three statements are equivalent:

(d) The norm $\|\cdot\|$ is Gâteaux-differentiable at x.

(e) $f_n - g_n \xrightarrow{*} 0$ in X^* whenever $f_n, g_n \in S_{X^*}$ satisfy $\lim_{n\to\infty} \langle x, f_n \rangle = \lim_{n\to\infty} \langle x, g_n \rangle = 1$.

(f) There is a unique $f \in S_{X^*}$ such that $\langle x, f \rangle = 1$.

Proof. (i) Suppose that (a) holds and let $\varepsilon > 0$. By Lemma 1.1.17, there exists $\delta > 0$ such that $\|x + h\| + \|x - h\| \leq 2 + \varepsilon \|h\|$ whenever $\|h\| < \delta$. Now suppose that $f_n, g_n \in S_{X^*}$ satisfy $\lim_{n\to\infty} \langle x, f_n \rangle = \lim_{n\to\infty} \langle x, g_n \rangle = 1$. Let $N \in \mathbb{N}$ be such that $\max\{|\langle x, f_n \rangle - 1|, |\langle x, g_n \rangle - 1|\} < \varepsilon \delta$ if $n \geq N$. Then if $n \geq N$ and $\|h\| < \delta$,

$$
\begin{aligned}
\mathrm{re}\,\langle h, f_n - g_n \rangle &= \mathrm{re}\{\langle x + h, f_n \rangle + \langle x - h, g_n \rangle - \langle x, f_n \rangle - \langle x, g_n \rangle\} \\
&\leq \|x + h\| + \|x - h\| - \mathrm{re}\,\langle x, f_n \rangle - \mathrm{re}\,\langle x, g_n \rangle \\
&\leq 2 + \varepsilon \|h\| - \mathrm{re}\,\langle x, f_n \rangle - \mathrm{re}\,\langle x, g_n \rangle \\
&\leq |\langle x, f_n \rangle - 1| + |\langle x, g_n \rangle - 1| + \varepsilon \|h\| < 3\varepsilon\delta.
\end{aligned}
$$

Thus if $n \geq N$,

$$
\begin{aligned}
\|f_n - g_n\| &= \sup_{\|h\|=1} |\langle h, f_n - g_n \rangle| = \sup_{\|h\|=1} |\mathrm{re}\,\langle h, f_n - g_n \rangle| \\
&= \sup_{\|h\|=1} |\mathrm{re}\,\langle \delta h, f_n - g_n \rangle| / \delta \leq 3\varepsilon.
\end{aligned}
$$

Hence $\lim_{n\to\infty} \|f_n - g_n\| = 0$; (b) holds.

On the other hand, if (a) is false, then by Lemma 1.1.17, there exist $\varepsilon > 0$ and a sequence (h_n) in X such that $\lim_{n\to\infty} \|h_n\| = 0$ and $\|x + h_n\| + \|x - h_n\| \geq 2 + \varepsilon \|h_n\|$ for all $n \in \mathbb{N}$. Given $n \in \mathbb{N}$, by the Hahn–Banach theorem there exist $f_n, g_n \in S_{X^*}$ such that $\langle x + h_n, f_n \rangle = \|x + h_n\|$ and $\langle x - h_n, g_n \rangle = \|x - h_n\|$. Plainly $\|x + h_n\| \to 1$ and $|\langle h_n, f_n \rangle| \leq \|h_n\| \to 0$. Thus

$$
\lim_{n\to\infty} \langle x, f_n \rangle = \lim_{n\to\infty} (\langle x + h_n, f_n \rangle - \langle h_n, f_n \rangle) = \lim_{n\to\infty} (\|x + h_n\| - \langle h_n, f_n \rangle) = 1,
$$

and in the same way it follows that $\lim_{n\to\infty} \langle x, g_n \rangle = 1$. Moreover,

$$
\begin{aligned}
|\langle h_n, f_n - g_n \rangle| &= |\langle x + h_n, f_n \rangle + \langle x - h_n, g_n \rangle - \langle x, f_n + g_n \rangle| \\
&\geq \|x + h_n\| + \|x - h_n\| - 2 \geq \varepsilon \|h_n\|.
\end{aligned}
$$

Hence $\|f_n - g_n\| \geq \varepsilon$ for all $n \in \mathbb{N}$, from which we have the equivalence of (a) and (b). That (c) is equivalent to (b) is clear.

Turning to (ii), similar arguments show that (d) and (e) are equivalent. Now suppose that (f) holds and let $f_n, g_n \in S_{X^*}$, $y \in S_X$ and $\varepsilon > 0$ be such that $\langle x, f_n \rangle \to 1, \langle x, g_n \rangle \to 1$ and $|\langle y, f_n - g_n \rangle| \geq \varepsilon$. The weak *-compactness of B_{X^*} implies that, by passage to a subsequence if necessary, there exist $f, g \in B_{X^*}$ to which $(f_n), (g_n)$ respectively converge in the weak* sense. Then $\langle x, f \rangle = \langle x, g \rangle = 1$ and $|\langle y, f - g \rangle| \geq \varepsilon$, so that $f \neq g$. This contradicts (f). The remaining implication is clear. $\square$

Proposition 1.1.20. *Let X be a Banach space. Then X^* is strictly convex if and only if $\|\cdot\|_X$ is Gâteaux-differentiable.*

Proof. From Proposition 1.1.3 we see that is strictly convex if and only if given any $x \in X \backslash \{0\}$, there is a unique $x^* \in S_{X^*}$ such that $\langle x, x^* \rangle = \|x\|$. Now use Lemma 1.1.19 (ii). $\square$

It turns out that uniform convexity of a space X has significant implications for the dual space X^*. To explain this we introduce a new object, the modulus of smoothness of a space.

Definition 1.1.21. Let X be a Banach space. The **modulus of smoothness** of X is the function $\rho_X : (0, \infty) \to [0, \infty)$ defined by

$$\rho_X(\tau) = \sup \left\{ \frac{\|x + \tau h\| + \|x - \tau h\|}{2} - 1 : x, h \in S_X \right\}.$$

If $\lim_{\tau \to 0} \rho_X(\tau)/\tau = 0$, the space X is said to be **uniformly smooth**.

Note that ρ_X really is a non-negative function since for all $x, h \in X$ we have $2 \|x\| = \|x + \tau h + x - \tau h\| \leq \|x + \tau h\| + \|x - \tau h\|$. Evidently the property of uniform smoothness is preserved on passage to a subspace. Some useful equivalents of uniform smoothness are given in the next lemma, the proof of which is omitted as it is quite similar to that of Lemma 1.1.17.

Lemma 1.1.22. *Let X be a Banach space. The following statements are equivalent:*

(i) *X is uniformly smooth.*

(ii) *The limit*

$$\lim_{t \to 0} \frac{\|x + \tau h\| - \|x\|}{t} = \langle h, \operatorname{grad} \|x\| \rangle$$

exists, uniformly for $x, h \in S_X$.

(iii) *The norm of X is Fréchet-differentiable on S_X and the map $x \longmapsto \operatorname{grad} \|x\| : S_X \to S_{X^*}$ is uniformly continuous.*

If $\|\cdot\|$ satisfies any of the equivalent conditions in the last lemma we shall say that it is **uniformly Fréchet-differentiable**.

A connection between the modulus of convexity of a space and the modulus of smoothness of its dual is provided by the following lemma.

Lemma 1.1.23. *Let X be a Banach space. Then for all $\tau > 0$,*

$$\rho_{X^*}(\tau) = \sup\left\{\frac{1}{2}\tau\varepsilon - \delta_X(\varepsilon) : 0 < \varepsilon \leq 2\right\}$$

and

$$\rho_X(\tau) = \sup\left\{\frac{1}{2}\tau\varepsilon - \delta_{X^*}(\varepsilon) : 0 < \varepsilon \leq 2\right\}.$$

Proof. Given $\varepsilon \in (0, 2]$ and $\tau > 0$, corresponding to any $x, y \in S_X$ such that $\|x - y\| \geq \varepsilon$ there are functionals $f, g \in S_{X^*}$ such that $\langle x + y, f \rangle = \|x + y\|$ and $\langle x - y, g \rangle = \|x - y\|$. Then

$$
\begin{aligned}
2\rho_{X^*}(\tau) &\geq \|f + \tau g\| + \|f - \tau g\| - 2 \\
&\geq \operatorname{re}\langle x, f + \tau g \rangle + \operatorname{re}\langle y, f - \tau g \rangle - 2 \\
&= \langle x + y, f \rangle + \tau \langle x - y, g \rangle - 2 \\
&= \|x + y\| + \tau \|x - y\| - 2.
\end{aligned}
$$

Hence $2 - \|x + y\| \geq \tau\varepsilon - 2\rho_{X^*}(\tau)$, so that $\delta_X(\varepsilon) + \rho_{X^*}(\tau) \geq \tau\varepsilon/2$. Thus $\rho_{X^*}(\tau) \geq \sup\left\{\frac{1}{2}\tau\varepsilon - \delta_X(\varepsilon) : 0 < \varepsilon \leq 2\right\}$.

For the reverse inequality, let $\tau > 0$ and $f, g \in S_{X^*}$. Given $\eta > 0$, there exist $x, y \in S_X$ such that $\operatorname{re}\langle x, f + \tau g \rangle \geq \|f + \tau g\| - \eta$ and $\operatorname{re}\langle x, f - \tau g \rangle \geq \|f - \tau g\| - \eta$. Hence

$$
\begin{aligned}
\|f + \tau g\| + \|f - \tau g\| - 2 &\leq \operatorname{re}\langle x + y, f \rangle + \tau \operatorname{re}\langle x - y, g \rangle - 2 + 2\eta \\
&\leq \|x + y\| - 2 + \tau \|x - y\| + 2\eta \\
&\leq -2\delta_X(\|x - y\|) + \tau \|x - y\| + 2\eta \\
&\leq 2\sup\left\{\frac{1}{2}\tau\varepsilon - \delta_{X^*}(\varepsilon) : 0 < \varepsilon \leq 2\right\} + 2\eta,
\end{aligned}
$$

from which the required inequality follows. The proof of the statement concerning $\rho_X(\tau)$ is similar. $\qquad\square$

Theorem 1.1.24. *Let X be a Banach space. Then:*

  (i) *X is uniformly convex if and only if X^* is uniformly smooth.*
  (ii) *X is uniformly smooth if and only if X^* is uniformly convex.*

Proof. (i) Let X be uniformly convex and let $\varepsilon_0 \in (0, 2]$. Then $\delta_X(\varepsilon) \geq \delta_X(\varepsilon_0) > 0$ whenever $\varepsilon \in [\varepsilon_0, 2]$. Let $\tau \in (0, \delta_X(\varepsilon_0))$. If $\varepsilon \in [\varepsilon_0, 2]$, then

$$\frac{1}{2}\varepsilon - \delta_X(\varepsilon)/\tau \leq \frac{1}{2}\varepsilon - \delta_X(\varepsilon_0)/\tau \leq \frac{1}{2}\varepsilon - 1 \leq 0.$$

Thus by Lemma 1.1.23,

$$\rho_{X^*}(\tau)/\tau = \sup_{0 < \varepsilon \leq \varepsilon_0}(\varepsilon/2 - \delta_X(\varepsilon)/\tau) \leq \sup_{0 < \varepsilon \leq \varepsilon_0}(\varepsilon/2) = \varepsilon_0/2.$$

It follows that $\lim_{\tau \to 0}\rho_X^*(\tau)/\tau = 0$: X^* is uniformly smooth.

Conversely, if X is not uniformly convex, then $\delta_X(\varepsilon_0) = 0$ for some $\varepsilon_0 \in (0, 2]$. By Lemma 1.1.23, if $\tau > 0$ then

$$\rho_{X^*}(\tau) = \sup_{0 < \varepsilon \leq 2} (\varepsilon\tau/2 - \delta_X(\varepsilon)) \geq \varepsilon_0\tau/2.$$

Hence $\limsup_{\tau \to 0} \rho_{X^*}(\tau)/\tau \geq \varepsilon_0/2$ and X^* is not uniformly smooth.

The proof of (ii) is similar, using the characterisation of $\rho_X(\tau)$ given in Lemma 1.1.23. $\qquad\square$

Note that from this result it follows that if X is uniformly smooth, it is reflexive.

For a Hilbert space H it is routine to verify that the modulus of smoothness is given by

$$\rho_H(\tau) = \left(1 + \tau^2\right)^{1/2} - 1,$$

so that H is uniformly smooth: of course this conclusion also follows from the fact that H is uniformly convex, taking Theorem 1.1.24 into account. It is known (see [69]) that for every Banach space X with dimension at least 2 we have

$$\delta_X(\varepsilon) \leq 1 - \left(1 - \varepsilon^2/4\right)^{1/2} \text{ and } \rho_X(\tau) \geq \left(1 + \tau^2\right)^{1/2} - 1.$$

This supports the view that Hilbert spaces are the 'most uniformly smooth' as well as the 'most uniformly convex' spaces.

To complete this section we deal with duality maps. A map $\mu : [0, \infty) \to [0, \infty)$ that is continuous, strictly increasing and satifies $\mu(0) = 0$, $\lim_{t \to \infty} \mu(t) = \infty$, is called a **gauge function**. A map J from a Banach space X to 2^{X^*}, the set of all subsets of X^*, is said to be a **duality map on X with gauge function μ** if for all $x \in X$,

$$J(x) = \{x^* \in X^* : \langle x, x^* \rangle = \|x^*\| \, \|x\| \, , \|x^*\| = \mu \, (\|x\|)\}.$$

By the Hahn–Banach theorem, for each $x \in X$ the set $J(x)$ is non-empty. It is also convex. To establish this, let $x^*, y^* \in J(x)$ and $\lambda \in (0, 1)$. Put $z^* = \lambda x^* + (1 - \lambda)y^*$ and observe that

$$\langle x, z^* \rangle = \lambda \langle x, x^* \rangle + (1 - \lambda) \langle x, y^* \rangle = \mu \, (\|x\|) \, \|x\| \, .$$

Hence $\|z^*\| \geq \mu \, (\|x\|)$. However,

$$\|z^*\| \leq \lambda \|x^*\| + (1 - \lambda) \|y^*\| = \mu \, (\|x\|)$$

and so $\|z^*\| = \mu \, (\|x\|)$. Thus $z^* \in J(x)$ and the proof is complete.

Let X be a Banach space with strictly convex dual X^* and let J be a duality map on X with gauge function μ. Then for each $x \in X$, the set $J(x)$ consists of precisely one point. In fact, for each $x \in X$, the points in $J(x)$ lie on the sphere in X^* with centre 0 and radius $\mu\left(\|x\|\right)$. If $J(x)$ contained two distinct points, the midpoint of the line segment joining them would be in the convex set $J(x)$, which is impossible as X^* is strictly convex.

In view of this result, we shall regard a duality map J on X as a map from X to X^* when X^* is strictly convex.

Proposition 1.1.25. *Let X be a Banach space with strictly convex dual X^* and let J be a duality map on X with gauge function μ. Then J is monotone in the sense that*

$$\operatorname{re}\langle x - y, Jx - Jy\rangle \geq 0 \text{ for all } x, y \in X.$$

If, in addition, X is strictly convex, then J is strictly monotone, by which is meant that

$$\operatorname{re}\langle x - y, Jx - Jy\rangle > 0 \text{ for all } x, y \in X \text{ with } x \neq y.$$

Proof. For all $x, y \subset X$,

$$\operatorname{re}\langle x - y, Jx - Jy\rangle = \mu\left(\|x\|\right)\|x\| + \mu\left(\|y\|\right)\|y\| - \operatorname{re}\langle x, Jy\rangle - \operatorname{re}\langle y, Jx\rangle$$
$$\geq \left(\mu\left(\|x\|\right) - \mu\left(\|y\|\right)\right)\left(\|x\| - \|y\|\right) \geq 0.$$

Now suppose that, in addition, X is strictly convex. We have to show that $x = y$ when $\operatorname{re}\langle x - y, Jx - Jy\rangle = 0$. From the argument above it is clear that $\|x\| = \|y\|$. Suppose that $x \neq y$. Then $x/\|x\| \neq y/\|y\|$ and so, by Proposition 1.1.3,

$$\|Jx\| = \langle x/\|x\|, Jx\rangle > \operatorname{re}\langle y/\|y\|, Jx\rangle,$$

so that $\operatorname{re}\langle y, Jx\rangle < \langle x, Jx\rangle$; similarly, $\operatorname{re}\langle x, Jy\rangle < \langle y, Jy\rangle$. Hence

$$0 = \operatorname{re}\langle x - y, Jx - Jy\rangle > \langle x, Jx\rangle + \langle y, Jy\rangle - \langle x, Jx\rangle - \langle y, Jy\rangle = 0,$$

and we have a contradiction. □

Proposition 1.1.26. *Let X be a Banach space with strictly convex dual X^* and let J be a duality map on X with gauge function μ; suppose that (x_n) is a sequence in X that converges to $x \in X$. Then $Jx_n \stackrel{*}{\rightharpoonup} Jx$. If, in addition, X^* is uniformly convex, then $Jx_n \to Jx$.*

Proof. It is enough to deal with the case in which $\mu(t) = t$ for all $t \geq 0$. We have to show that for each $u \in X$, $\langle u, Jx_n\rangle \to \langle u, Jx\rangle$. Since the sequence (x_n) and the points x, u span a separable subspace of X, we may assume without loss of generality that X is separable. But then, as $(\|Jx_n\|)$ is bounded, it follows (see, for example, [83], Theorem 4.41-A) that, by passage to a subsequence if necessary, (Jx_n) converges in the weak*-topology of X^*, to x^*, say. Hence

$$\langle x_n, Jx_n\rangle \to \langle x, x^*\rangle, \quad \|x^*\| \leq \liminf_{n\to\infty}\|Jx_n\| = \lim_{n\to\infty}\|x_n\| = \|x\|,$$

and $\langle x_n, Jx_n \rangle = \|x_n\|^2 \to \|x\|^2$. The definition of J thus ensures that $x^* = Jx$. This argument holds for every weak $*$-convergent subsequence of (Jx_n), and so the proof of the first part is complete.

If X^* is uniformly convex, then since $Jx_n \rightharpoonup Jx$ (note that X^* is reflexive and hence so is X) and $\|Jx_n\| = \|x_n\| \to \|x\| = \|Jx\|$, we have $Jx_n \to Jx$. $\qquad \square$

When X is reflexive and X, X^* are strictly convex, it turns out that duality maps are surjective. This is a consequence of a general result concerning monotone maps which, although well-known (see, for example, [12]), we prove here for the convenience of the reader.

Theorem 1.1.27. *Let X be a reflexive Banach space and let $T : X \to X^*$ be such that*

(i) $\mathrm{re}\, \langle x - y, Tx - Ty \rangle \geq 0$ *for all* $x, y \in X$,
(ii) $Tx_n \overset{*}{\rightharpoonup} Tx$ *whenever* $x_n \to x$.
(iii) $\mathrm{re}\, \langle x, Tx \rangle / \|x\| \to \infty$ *when* $\|x\| \to \infty$.

Then T is surjective.

Proof. For simplicity of exposition we take X to be real: the proof in the complex case proceeds along similar lines.

Step. 1. We claim that if there exist $x_0 \in X$ and $y_0^* \in X^*$ such that

$$\langle x - x_0, Tx - y_0^* \rangle \geq 0 \text{ for all } x \in X, \qquad (1.1.3)$$

then $Tx_0 = y_0^*$. To establish this, let $y \in X$ and put $x_t = x_0 + ty$ $(t > 0)$. Then in view of (1.1.3), $\langle y, Tx_t - y_0^* \rangle \geq 0$, so that by property (ii) above, $\langle y, Tx_0 - y_0^* \rangle \geq 0$. As this holds for all $y \in X$, the claim follows.

Step 2. We assert that the theorem holds when $\dim X < \infty$, in which case the weak $*$-convergence of (ii) can be replaced by strong convergence. We may suppose that X is an n-dimensional Hilbert space with $X^* = X$; let $I : X \to X$ be the identity map and put $U = I - T$. Note that $y \in T(X)$ if and only if $0 \in T_y(X)$, where $T_y x = Tx - y$ $(x \in X)$; moreover, it is easy to see that T_y has properties (i)–(iii). It is therefore enough to show that $0 \in T(X)$, or equivalently that U has a fixed point. Observe also that by (iii),

$$\langle x, Ux \rangle = \|x\|^2 - \langle x, Tx \rangle \leq \|x\|^2$$

for large enough $\|x\|$, say $\|x\| \geq R > 0$. Define $V : X \to X$ by

$$Vx = \begin{cases} Ux & \text{if} & \|Ux\| \leq R, \\ Ux/\|Ux\| & \text{if} & \|Ux\| > R. \end{cases}$$

Plainly V maps the closed ball B_R in X (centred at the origin and with radius R) continuously into itself, and so, by Brouwer's fixed point theorem, the restriction of V to B_R has a fixed point, say x_0. If $\|x_0\| < R$, then $x_0 = Vx_0 = Ux_0$ and x_0 is a fixed point of U. If $\|x_0\| = R$, then $Vx_0 = \lambda Ux_0 = x_0$, where $\lambda = R/\|Ux_0\| \le 1$; and as

$$\langle x_0, x_0 \rangle = \lambda \langle x_0, Ux_0 \rangle \le \lambda \|x_0\|^2,$$

we must have $\lambda = 1$ and again x_0 is a fixed point of U.

Now that these two steps have been completed we can give the proof of the theorem. As before it is enough to show that $0 \in T(X)$. First note that condition (iii) can be written as

$$\langle x, Tx \rangle \ge c(\|x\|)\|x\|,$$

where

$$c(r) := \inf_{\|x\|=r} \langle x, Tx \rangle / \|x\| \to \infty \text{ as } r \to \infty.$$

Let Λ be the directed set of all finite-dimensional linear subspaces of X, ordered by inclusion. For each $F \in \Lambda$ let j_F be the embedding of F in X and j_F^* the adjoint projection of X^* onto F^*; set $T_F = j_F^* \circ T \circ j_F : F \to F^*$. Then T_F is continuous, and for all $x, y \in F$,

$$\langle y, T_F x \rangle = \langle y, j_F^* Tx \rangle = \langle j_F y, Tx \rangle = \langle y, Tx \rangle,$$

which implies that

$$\langle x - y, T_F x - T_F y \rangle = \langle x - y, Tx - Ty \rangle \ge 0$$

and

$$\langle x, T_F x \rangle = \langle x, Tx \rangle \ge c(\|x\|)\|x\|.$$

Hence T_F satisfies the same hypotheses as T, but on the finite-dimensional space F instead of X. By Step 2, there exists $x_F \in F$ such that $T_F x_F = 0$. Since

$$0 = \langle x_F, T_F x_F \rangle \ge c(\|x_F\|)\|x_F\|$$

and $c(r) \to \infty$ as $r \to \infty$, it follows that there exists M such that for all $F \in \Lambda$, $\|x_F\| \le M$. Given any $F_0 \in \Lambda$, put

$$V_{F_0} = \bigcup_{F \in \Lambda, F_0 \subset F} \{x_F\} \subset B_M,$$

where B_M is the closed ball in X with centre 0 and radius M, and let $\overline{V_{F_0}}$ be the weak closure of V_{F_0}. Since X is reflexive, B_M is weakly compact; and as the family of all $\overline{V_{F_0}}$ plainly has the finite intersection property, the set $\cap_{F_0 \in \Lambda} \overline{V_{F_0}}$ is non-empty. Let x_0 be an element of this set.

We claim that $Tx_0 = 0$. Let $x \in X$, let $F_0 \in \Lambda$ be such that $x_0 \in F_0$ and let $x_F \in V_{F_0}$. As T is monotone, $\langle x_F - x, Tx_F - Tx \rangle \ge 0$; moreover, $\langle x_F - x, Tx_F \rangle = \langle x_F - x, T_F x_F \rangle = 0$. Hence $\langle x - x_F, Tx \rangle \ge 0$, which implies that $\langle x - x_0, Tx \rangle \ge 0$ for all $x \in X$. Step 1 now shows that $Tx_0 = 0$ and completes the proof of the theorem. $\qquad\square$

Corollary 1.1.28. *Let X be a reflexive Banach space such that X and X^* are strictly convex, and let $J : X \to X^*$ be a duality map with gauge function μ. Then J maps X surjectively onto X^*, and the map $f \longmapsto J^{-1}(f)$ is a duality map on X^* with gauge function μ^{-1} (X^{**} being identified with X).*

Proof. That J is surjective follows immediately from Proposition 1.1.25 and Theorem 1.1.27. As for the second part, by the definition of J we have $\langle x, Jx \rangle = \|Jx\| \, \|x\|$, $\|Jx\| = \mu (\|x\|)$ for all $x \in X$. Given any $f \in X^*$ let x be the unique element of X such that $Jx = f$. Then

$$\langle J^{-1}f, f \rangle = \langle x, Jx \rangle = \|Jx\| \, \|x\| = \|f\| \, \|J^{-1}f\|,$$

$$\|f\| = \|Jx\| = \mu (\|x\|) = \mu \left(\|J^{-1}f\| \right), \, \|J^{-1}f\| = \mu^{-1} (\|f\|),$$

and so J^{-1} is the required duality map. $\qquad\square$

When X is a Hilbert space, so that X^* may be identified with X, the most natural duality map on X is the identity map, corresponding to the gauge function μ with $\mu(t) = t$. When $1 < p < \infty$ and $\mu(t) = t^{p-1}$ ($t \geq 0$), it is easy to check that the duality map J on $L_p(\Omega)$ (where Ω is, for example, an open subset of $\mathbb{R}^n$) is given by $J(u) = |u|^{p-2} u$; the duality map on l_p with the same gauge function is defined by $J((x_k)) = \left(|x_k|^{p-2} x_k \right)$.

Definition 1.1.29. Suppose that X is a Banach space with Gâteaux-differentiable norm. Let $(\cdot, \cdot)_X : X \times X \to \mathbb{K}$ ($\mathbb{R}$ or $\mathbb{C}$) be defined by

$$(x, h)_X = \begin{cases} \|x\| \, \langle h, \operatorname{grad} \|x\| \rangle_X, & x, h \in X, x \neq 0, \\ 0, & x = 0, h \in X. \end{cases}$$

We refer to $(x, h)_X$ as the **semi-inner product** of x and h.

Note that $(x, h)_X$ depends linearly on h and that $(x, x)_X = \|x\|^2$; in general, $(x, h)_X \neq (h, x)_X$. Of course, the semi-inner product above can be expressed in terms of the duality map J on the space X with gauge function μ. In fact, it is easy to verify that

$$(x, h)_X = \frac{\|x\|}{\mu (\|x\|)} \langle h, Jx \rangle_X \quad (x \neq 0).$$

If X is a Hilbert space, $(\cdot, \cdot)_X$ is simply the inner product on X. When $X = L_p(\Omega)$, $p \in (1, \infty)$ and $\mu(t) = t^{p-1}$ it can be checked that

$$(x, h)_X = \|x\|^{2-p} \int_\Omega |x(\xi)|^{p-2} \, \overline{x(\xi)} h(\xi) d\xi.$$

In [52], James introduced the following notion of orthogonality.

Definition 1.1.30. Let X be a Banach space with a strictly convex dual. We say that $x \in X$ is *j*-**orthogonal** (or orthogonal in the sense of James or Birkhoff) to $y \in X$, and write $x \perp^j y$, if

$$\|x\|_X \leq \|x + ty\|_X \quad \text{for every } t \in \mathbb{R}.$$

An element $x \in X$ is j-orthogonal to a set $M \subset X$ ($x \perp^j M$) if x is j-orthogonal to every element in M. A set $M_1 \subset X$ is j-orthogonal to $M_2 \subset X$ if every $x \in M_1$ is j-orthogonal to every $y \in M_2$.

We note that j-orthogonality is not symmetric on non-Hilbert spaces, that is, $x \perp^j y$ does not imply that $y \perp^j x$; in fact it is shown in [52] that if X is strictly convex with $\dim X \geq 3$, then the symmetry of j-orthogonality implies that X is an inner-product space. The linkage between j-orthogonality and the semi-inner product of Definition 1.1.29 is given in the following proposition proved in [52]:

Proposition 1.1.31. *Let X be a Banach space with a strictly convex dual and $x, h \in X$. Then $x \perp^j h$ if and only if $(x,h)_X := \|x\| \langle h, \mathrm{grad}\, \|x\| \rangle_X = 0$. Moreover, given $x \in X \setminus \{0\}$ and $y \in X$, there is precisely one $\lambda \in \mathbb{R}$ such that $x \perp^j \lambda x + y$ and this is given by*

$$\langle y, \mathrm{grad}\, \|x\| \rangle_X = -\lambda \|x\|.$$

Proof. First suppose that $x \perp^j h$, $x \neq 0$. Then

$$(x, h)_X = \|x\| \langle h, \mathrm{grad}\, \|x\| \rangle_X = \lim_{t \to 0} \frac{\|x + th\| - \|x\|}{t}.$$

Since $x \perp^j h$,

$$\frac{\|x + th\| - \|x\|}{t} \geq 0 \text{ when } t > 0, \quad \text{and} \quad \frac{\|x + th\| - \|x\|}{t} \leq 0 \text{ when } t < 0.$$

Hence $(x, h)_X = 0$. This plainly also holds when $x = 0$.

Conversely, suppose that $(x, h)_X = 0$, so that $\lim_{t \to 0} t^{-1} (\|x + th\| - \|x\|) = 0$. Since $t \longmapsto t^{-1} (\|x + th\| - \|x\|)$ is monotonic increasing in $t > 0$, and hence (by consideration of $-h$ in place of h) $t \longmapsto t^{-1} (\|x + th\| - \|x\|)$ is monotonic decreasing in $t < 0$, we see that $\|x + th\| \geq \|x\|$ for all t. Thus $x \perp^j h$.

Finally, if $x, y \in X$, $x \neq 0$, then $x \perp^j (\lambda x + y)$ if and only if

$$\langle \lambda x + y, \mathrm{grad}\, \|x\| \rangle_X = 0,$$

which holds if and only if

$$\langle y, \mathrm{grad}\, \|x\| \rangle_X = -\lambda \langle x, \mathrm{grad}\, \|x\| \rangle_X = -\lambda \|x\|. \qquad \square$$

An important part is played in the theory to follow by the notion of a **polar set**: given any linear subspace N of a Banach space X, the polar set of N is the linear subspace of X^* given by

$$N^0 := \{x^* \in X^* : \langle x, x^* \rangle_X = 0 \text{ for all } x \in N\}.$$

Clearly, N^0 is a closed subspace of X^* and $N^0 = (\overline{N})^0$. Similarly, if M is a linear subspace of X^*, we set

$$^0M := \{x \in X : \langle x, x^* \rangle_X = 0 \text{ for all } x^* \in M\},$$

^0M being closed and $^0M =^0 (\overline{M})$. We also have that $\overline{N} =^0 (N^0)$.

A decomposition of X in terms of James orthogonality was given by Alber in [4].

Definition 1.1.32. Let M_1, M_2 be closed subsets of X. We say that X is the **James orthogonal direct sum** of M_1 and M_2, written $X = M_1 \uplus M_2$, if:

1. for each $x \in X$ there exists a unique decomposition $x = m_1 + m_2$, where $m_1 \in M_1, m_2 \in M_2$,

2. $M_2 \perp^j M_1$,

3. $M_1 \cap M_2 = \{0\}$.

Alber's theorem is

Theorem 1.1.33. *Let X be uniformly convex and uniformly smooth and let M be a closed linear subspace of X with polar $M^0 \subset X^*$. Let J_X be a duality map, normalised in the sense that it has gauge function $\mu(t) = t$ for all $t \geq 0$. Then*

$$X = M \uplus J_X^{-1} M^0 \quad \text{and} \quad X^* = M^0 \uplus J_X M.$$

1.2 Bases

Let X be an infinite-dimensional normed linear space with norm $\|\cdot\|$. A sequence $(x_n)_{n \in \mathbb{N}}$ of elements of X is called a (**Schauder**) **basis** of X if, given any $x \in X$, there is a unique sequence $(a_n)_{n \in \mathbb{N}}$ of scalars such that

$$x = \sum_{n=1}^{\infty} a_n x_n; \text{ that is, } \lim_{N \to \infty} \left\| x - \sum_{n=1}^{N} a_n x_n \right\| = 0.$$

It is called a **basic sequence** if it is a basis of the closed linear span of the x_n. If X has a basis (x_n), the set $\{x_n : n \in \mathbb{N}\}$ is plainly linearly independent, and X is separable. Given a basis $(x_n)_{n \in \mathbb{N}}$ of X and $N \in \mathbb{N}$, we define a map $P_N : X \to X$ by

$$P_N(x) = \sum_{n=1}^{N} a_n x_n, \ x = \sum_{n=1}^{\infty} a_n x_n \in X.$$

These linear maps P_N are often called the **canonical projections** associated with the given basis; some of their fundamental properties are given in the next lemma.

Lemma 1.2.1. *Let X be a normed linear space with Schauder basis $(x_n)_{n\in\mathbb{N}}$. Then:*

  (i) *for each $n \in \mathbb{N}$, $\dim(P_n(X)) = n$;*
  (ii) *for every $m, n \in \mathbb{N}$, $P_m P_n = P_n P_m = P_{\min\{m,n\}}$;*
  (iii) *for every $x \in X$, $P_n x \to x$ in X.*

Conversely, if there are bounded linear projections Q_n $(n \in \mathbb{N})$ on a normed linear space X that have properties (i)–(iii)*, then they are the canonical projections associated with some basis of X.*

Proof. Since $\{x_n : n \in \mathbb{N}\}$ is linearly independent, (i) follows; (ii) and (iii) are obvious. For the converse, let Q_0 be the zero operator and choose $e_i \in Q_i(X) \cap \ker(Q_{i-1})$ $(i \in \mathbb{N})$. Then for all $x \in X$,

$$x = \lim_{n\to\infty} (Q_n(x) - Q_0(x)) = \lim_{n\to\infty} \sum_{i=1}^{n} (Q_i(x) - Q_{i-1}(x)) = \sum_{i=1}^{\infty} \alpha_i e_i$$

for some scalars α_i, since $\dim(Q_i(X)/Q_{i-1}(X)) = 1$. The α_i are uniquely determined, for if $x = \sum_{i=1}^{\infty} \beta_i e_i$, then as Q_n is continuous,

$$Q_n x = \lim_{m\to\infty} \sum_{i=1}^{m} \beta_i Q_n e_i = \sum_{i=1}^{n} \beta_i Q_n e_i$$

since by (ii), $Q_n e_i = 0$ if $i > n$. By (ii) again, if $i \leq n$,

$$Q_n e_i = Q_n Q_i e_i = Q_i e_i = e_i;$$

hence $Q_n x = \sum_{i=1}^{n} \beta_i e_i$, so that $\beta_i e_i = Q_i x - Q_{i-1} x = \alpha_i e_i$. $\square$

Our next aim is to show that the canonical projections have uniformly bounded norms. The strategy is to show that this is so with respect to another norm on X, and then to prove that this norm is equivalent to the original norm on X. First we need the following result concerning the extension of bases.

Lemma 1.2.2. *Let $(x_n)_{n\in\mathbb{N}}$ be a Schauder basis of a normed linear space X with corresponding canonical projections P_n, and suppose that $\sup_{n\in\mathbb{N}} \|P_n\| < \infty$. Then $(x_n)_{n\in\mathbb{N}}$ is a Schauder basis of the completion $\widetilde{X}$ of X. (It is assumed that X is identified with a dense subset of $\widetilde{X}$.)*

Proof. For each $n \in \mathbb{N}$ let $\widetilde{P}_n$ be the extension of P_n to $\widetilde{X}$. We claim that the $\widetilde{P}_n$ have properties (i)–(iii) of Lemma 1.2.1. Since $P_n(X)$ is finite-dimensional, it is closed in $\widetilde{X}$, so that $\widetilde{P}_n(\widetilde{X}) = P_n(X)$: (i) follows. The continuity of the P_n gives (ii). As $P_n x \to x$ for all x in the dense subset X of $\widetilde{X}$ and the P_n are uniformly bounded, it follows that $\widetilde{P}_n x \to x$ for all $x \in \widetilde{X}$, and hence (iii) holds. Since $x_n \in P_n(X) \cap \ker P_{n-1}$, we have $x_n \in \widetilde{P}_n(\widetilde{X}) \cap \ker \widetilde{P}_{n-1}$ for all n: the $\widetilde{P}_n$ are thus the canonical projections associated with the Schauder basis $(x_n)_{n\in\mathbb{N}}$ of $\widetilde{X}$. $\square$

Theorem 1.2.3. *Let* $(x_n)_{n\in\mathbb{N}}$ *be a Schauder basis of a Banach space* X, *with associated canonical projections* P_n. *Then* $\sup_{n\in\mathbb{N}}\|P_n\| < \infty$.

Proof. The result would follow immediately from the principle of uniform boundedness if the P_n were known to be continuous. In the absence of this a priori knowledge we proceed as follows. Define $\|\cdot\|_1$ on X by

$$\|x\|_1 = \sup_{n\in\mathbb{N}}\left\|\sum_{i=1}^{n} a_i x_i\right\|, \, x = \sum_{i=1}^{\infty} a_i x_i.$$

Claim 1. $\|\cdot\|_1$ is a norm on X.

The triangle inequality and homogeneity are easily verified; and since $\|x\| = \lim_{n\to\infty}\|\sum_{i=1}^{n} a_i x_i\| \le \|x\|_1 < \infty$ for all $x \in X$, the claim follows.

Claim 2. $(x_n)_{n\in\mathbb{N}}$ is a Schauder basis of $(X, \|\cdot\|_1)$ and $\|P_n\|_1 \le 1$ for all $n \in \mathbb{N}$.

To establish this, the converse part of Lemma 1.2.1 is used. Properties (i) and (ii) are easily checked. As for (iii), if $x \in X$, then

$$\|x - P_m x\|_1 = \sup_{n\in\mathbb{N}}\|P_n x - P_n P_m x\| = \sup_{n\ge m}\|P_n x - P_m x\| \to 0$$

as $m \to \infty$. Hence $(x_n)_{n\in\mathbb{N}}$ is a Schauder basis of $(X, \|\cdot\|_1)$. Moreover, for all $m \in \mathbb{N}$,

$$\|P_m\|_1 = \sup_{\|x\|_1\le 1}\|P_m x\|_1 = \sup_{\|x\|_1\le 1}\sup_{n\in\mathbb{N}}\|P_n P_m x\| = \sup_{n\in\mathbb{N}}\sup_{\|x\|_1\le 1}\|P_n P_m x\|$$

$$= \sup_{n\in\mathbb{N}}\sup\left\{\|P_n P_m x\| : \sup_{i}\|P_i x\| \le 1\right\} \le 1.$$

Claim 3. $(x_n)_{n\in\mathbb{N}}$ is a Schauder basis of the completion X_1 of $(X, \|\cdot\|_1)$.

This is just what Lemma 1.2.2 asserts.

Claim 4. $(X, \|\cdot\|_1)$ is complete.

By Claim 3, $(x_n)_{n\in\mathbb{N}}$ is a Schauder basis of X_1. Given $x_1 \in X_1$, there is a unique sequence (α_i) of scalars such that $x_1 = \sum \alpha_i x_i$, convergence being with respect to $\|\cdot\|_1$. Since $\|x\| \le \|x\|_1$ if $x \in X$, the series $\sum \alpha_i x_i$ also converges in the sense of the norm $\|\cdot\|$, to some point $x \in X$. As shown in the proof of Claim 2, verification of property (iii), $\sum \alpha_i x_i$ must converge to x in $(X, \|\cdot\|_1)$. Hence $x_1 = x \in X$.

Claim 5. $\|\cdot\|_1$ and $\|\cdot\|$ are equivalent norms on X.

For since the identity map $id : (X, \|\cdot\|_1) \to (X, \|\cdot\|)$ is a continuous linear bijection, it follows from the inverse mapping theorem that id^{-1} is continuous.

The proof of the theorem is now completed by observing that $\|P_n\|_1 \le 1$ for all $n \in \mathbb{N}$ and using Claim 5. $\qquad\square$

The number $\sup_{n\in\mathbb{N}}\|P_n\|$ is called the **basis constant** of the basis $(x_n)_{n\in\mathbb{N}}$ and is often denoted by $bc(x_n)$. If $\|P_n\| = 1$ for all $n \in \mathbb{N}$, the basis (x_n) is said to be **monotone**.

If X is a Hilbert space, a basis $(x_n)_{n\in\mathbb{N}}$ of X is called a **Riesz basis** if the map $(a_n) \longmapsto (\sum_{n=1}^{\infty} a_n x_n)$ is an isomorphism of l_2 onto X. This means that there are positive constants c, C such that for all $(a_n) \in l_2$,

$$c\sum_{n=1}^{\infty} |a_n|^2 \le \left\|\sum_{n=1}^{\infty} a_n x_n\right\|^2 \le C\sum_{n=1}^{\infty} |a_n|^2.$$

It is plain that any complete orthonormal system in a separable Hilbert space is a Riesz basis. Examples of such systems are the sequence of trigonometric functions $(e^{in\pi x})_{n\in\mathbb{Q}}$ in the complex space $L_2(-1,1)$ and the sequence of standard unit vectors in l_2.

Outside Hilbert spaces more effort is often needed to show that given sets of functions form a basis. We illustrate this by the following examples.

(i) When $1 < p < \infty$, a basis of $L_p(-1,1)$ is given by $(e^{in\pi x})_{n\in\mathbb{Z}}$. This follows from an important result due to M. Riesz, namely that (see [43], 12.10.1)

$$\lim_{N\to\infty} \left\| f - \sum_{|n|\le N} f_n e^{in\pi x} \right\|_p = 0, \tag{1.2.1}$$

for all $f \in L_p(-1,1)$, where $f_n = \frac{1}{2}\int_{-1}^{1} f(x) e^{-in\pi x} dx$ and $\|\cdot\|_p$ is the usual norm on $L_p(-1,1)$; when $p = 1$, (1.2.1) is false. Given any $f \in L_p(0,1)$, its odd extension to $L_p(-1,1)$ has a unique representation in terms of the $\sin n\pi x$, which means that $(\sin n\pi x)_{n\in\mathbb{N}}$ is a basis of $L_p(0,1)$ when $1 < p < \infty$; a similar argument applies to $(\cos n\pi x)_{n\in\mathbb{N}_0}$.

(ii) The Haar functions provide another example of a basis. These are the functions $h_n : [0,1] \to \mathbb{R}$ $(n \in \mathbb{N}_0)$ defined by

$$h_0(t) = 1,$$

and, with each $n \in \mathbb{N}$ represented as $n = 2^j + k$ for some $j \in \mathbb{N}_0$ and $k \in \{0, 1, \ldots, 2^{j-1}\}$,

$$h_n(t) = \begin{cases} -2^{j/2}, & 2^{-j}k \le t < 2^{-j-1}(2k+1), \\ 2^{j/2}, & 2^{-j-1}(2k+1) \le t < 2^{-j}(k+1), \\ 0, & \text{otherwise.} \end{cases}$$

To show that these form a basis of $L_p(0,1)$ when $1 \le p < \infty$ we proceed as follows. For each pair (j,k) with $j \in \mathbb{N}_0$ and $k \in \{0, 1, \ldots, 2^j - 1\}$ put $N(j,k) = 2^j + k$, let $S_{N(j,k)}$ be the collection of all intervals that are either of the form

$(2^{-j-1}s, 2^{-j-1}(s+1))$ $(s = 0, 1, \ldots, 2k+1)$ or $(2^{-j}s, 2^{-j}(s+1))$ $(s = k+1,$ $\ldots, 2^j - 1)$, and let $F_{N(j,k)}$ be the space of all functions that are constant on the intervals in $S_{N(j,k)}$. Then $\dim F_{N(j,k)} = N(j,k) + 1$; and as $h_n \in F_{N(j,k)}$ for $n \in \{0, 1, \ldots, N(j,k)\}$ it follows that $\mathrm{sp}\{h_0, \ldots, h_{N(j,k)}\} = F_{N(j,k)}$ and hence the span of the h_n $(n \in \mathbb{N}_0)$ is dense in $L_p(0,1)$. For each $f \in L_p(0,1)$ define

$$P_{N(j,k)}f = \sum_{I \in S_{N(j,k)}} |I|^{-1} \left(\int_I f(x)dx \right) \chi_I;$$

this is a partial sum projection and by Hölder's inequality

$$\|P_{N(j,k)}f\|_p^p = \sum_{I \in S_{N(j,k)}} |I|^{-p} \left| \int_I f(x)dx \right|^p \int_0^1 \chi_I \leq \sum_{I \in S_{N(j,k)}} |I|^{1-p+p/p'} \int |f|^p$$

$$= \sum_{I \in S_{N(j,k)}} \int |f|^p = \|f\|_p^p.$$

Thus $\|P_n\|_p = 1$ for all $n \in \mathbb{N}_0$, and the basis property now results from Lemma 1.2.1.

(iii) If $1 \leq p < \infty$, $n \in \mathbb{N}$ and Ω is any measurable subset of $\mathbb{R}^n$, then $L_p(\Omega)$ has a basis. This has already been established when $n = 1$ and $\Omega = (0,1)$, and the claim follows from the fact that $L_p(\Omega)$ is isometrically isomorphic to $L_p(0,1)$. This fact is given in [85], p. 236; for the convenience of the reader we indicate briefly how a proof may be constructed. Let μ_n be Lebesgue n-measure on Ω and let $f \in L_1(\Omega)$ be a positive function with $\|f\|_1 = 1$. Let ν be the measure on Ω whose Radon–Nikodym derivative with respect to μ_n is f: ν is a probability measure since $\nu(\Omega) = \int_\Omega f(x)dx = 1$. Define $T : L_p(\Omega) \to L_p(\Omega, \nu)$ by $Tg = f^{-1/p}g$ $(g \in L_p(\Omega))$. Then

$$\|Tg \mid L_p(\Omega, \nu)\|^p = \int_\Omega |Tg|^p d\nu = \int_\Omega |g|^p dx,$$

and so T is an isometry of $L_p(\Omega)$ onto $L_p(\Omega, \nu)$. It is therefore enough to show that $L_p(\Omega, \nu)$ is isometrically isomorphic to $L_p(0,1)$. Let S be the family of all Borel-measurable subsets of Ω and let $(B_m)_{m \in \mathbb{N}} \subset S$ be a sequence that generates S. An inductive argument shows that for each $m \in \mathbb{N}_0$ there is a finite S-partition $P_m = \left(P_1^{(m)}, P_2^{(m)}, \ldots, P_{k(m)}^{(m)} \right)$ of Ω such that

(a) $\{B_1, \ldots, B_m\}$ is contained in the σ-algebra generated by P_m;

(b) $\nu\left(P_i^{(m)} \right) \leq 2^{-m}$ for $i \in \{1, 2, \ldots, k_m\}$;

(c) P_m is a subpartition of P_{m-1} if $m > 1$: for each $i \in \{1, 2, \ldots, k_{m-1}\}$ there are $s_m(i), t_m(i) \in \{1, 2, \ldots, k_m\}$, with $s_m(i) \le t_m(i)$, such that

$$P_i^{(m-1)} = \bigcup_{j=s_m(i)}^{t_m(i)} P_j^{(m)}.$$

For each $m \in \mathbb{N}$ and $i \in \{1, 2, \ldots, k_m\}$ put

$$\widetilde{P}_i^{(m)} = \left[\sum_{j \le i-1} \nu\left(P_j^{(m)}\right), \sum_{j \le i} \nu\left(P_j^{(m)}\right) \right)$$

if $i < k_m$, and

$$\widetilde{P}_{k_m}^{(m)} = \left[\sum_{j \le k_m-1} \nu\left(P_j^{(m)}\right), \sum_{j \le k_m} \nu\left(P_j^{(m)}\right) \right].$$

Set $\widetilde{P}^{(m)} = \left(\widetilde{P}_1^{(m)}, \widetilde{P}_2^{(m)}, \ldots, \widetilde{P}_{k_m}^{(m)} \right)$. Then $\widetilde{P}^{(m)}$ is a Borel partition of $[0, 1]$, with $\mu_1\left(\widetilde{P}_i^{(m)}\right) = \nu\left(P_i^{(m)}\right)$ for each $i \le k_m$, and $\cup_{m \in \mathbb{N}}\widetilde{P}^{(m)}$ generates the Borel σ-algebra on $[0, 1]$. Now let, for each $m \in \mathbb{N}$, V_m and $\widetilde{V}_m$ be the linear spans of $\chi_{P_1^{(m)}}, \ldots, \chi_{P_{k_m}^{(m)}}$ and $\chi_{\widetilde{P}_1^{(m)}}, \ldots, \chi_{\widetilde{P}_{k_m}^{(m)}}$, respectively. Then $V := \cup_{m=1}^{\infty} V_m$ and $\widetilde{V} := \cup_{m=1}^{\infty} \widetilde{V}_m$ are dense subspaces of $L_p(\Omega, \nu)$ and $L_p(0, 1)$, respectively. The map $A : V \to \widetilde{V}$ given by

$$A\left(\sum_{i=1}^{k_m} a_i \chi_{P_i^{(m)}} \right) = \sum_{i=1}^{k_m} a_i \chi_{\widetilde{P}_i^{(m)}}$$

is an isometry with range dense in $L_p(0, 1)$ and can therefore be extended to an isometry from $L_p(\Omega, \nu)$ onto $L_p(0, 1)$. Hence $L_p(\Omega)$ is isometrically isomorphic to $L_p(0, 1)$, and so, by Proposition 1.2.16 below, $L_p(\Omega)$ has a basis. This argument (which follows the presentation given in the Texas A&M University lecture notes of T. Schlumprecht) also applies to any space $L_p(\Lambda, \Sigma, \mu)$, where (Λ, Σ, μ) is a separable measure space (in the sense that Σ is generated by a countable family of subsets of Λ) and μ is a measure with no atoms. Note that the existence of a basis in much more general spaces than L_p is established in [85], Chapters 1 and 2.

In contrast to these non-obvious results, it is trivial that a basis in the sequence space l_p ($1 \le p < \infty$) is given by the standard unit vectors.

Turning for the moment to basic sequences, we observe that they may be characterised as follows.

Proposition 1.2.4. *A sequence $(x_n)_{n \in \mathbb{N}}$ in a Banach space X is a basic sequence if and only if there exists a constant $K > 0$ such that for all $m, n \in \mathbb{N}$ with $m < n$*

and all scalars $a_1, \ldots, a_n$,

$$\left\| \sum_{i=1}^{m} a_i x_i \right\| \leq K \left\| \sum_{i=1}^{n} a_i x_i \right\|.$$

If such a constant exists, the smallest K is $bc(x_n)$.

Proof. Suppose that $(x_n)_{n \in \mathbb{N}}$ is a basic sequence with associated projections P_n. Then if $m < n$,

$$\left\| \sum_{i=1}^{m} a_i x_i \right\| = \left\| P_m \left(\sum_{i=1}^{n} a_i x_i \right) \right\| \leq \| P_m \| \left\| \sum_{i=1}^{n} a_i x_i \right\| \leq bc(x_n) \left\| \sum_{i=1}^{n} a_i x_i \right\|.$$

Conversely, if the given condition holds, define projections P_m on the linear span of the x_i by

$$P_m \left(\sum_{i=1}^{n} a_i x_i \right) = \sum_{i=1}^{m} a_i x_i$$

for $m < n$. Then $\| P_m \| \leq K$. Moreover, the P_m satisfy conditions (i)–(iii) of Lemma 1.2.1 on $\mathrm{sp}\{x_i\}$, and so, in view of Lemma 1.2.2, (x_n) is a basis of $\overline{\mathrm{sp}}\{x_i\}$ and $bc(x_n) \leq K$. $\qquad \square$

To give further information about the existence of basic sequences it is convenient first to establish the following lemma.

Lemma 1.2.5. *Let B be a finite-dimensional subspace of an infinite-dimensional Banach space X and let $\varepsilon > 0$. Then there exists $x \in S_X$ such that $\|y\| \leq (1 + \varepsilon) \|y + \lambda x\|$ for every $y \in B$ and every scalar λ.*

Proof. Obviously it may be supposed that $\varepsilon \in (0, 1)$. Since $\dim B < \infty$, there exist $y_1, \ldots, y_m \in B \cap S_X$ such that given any $y \in B \cap S_X$, $\|y - y_i\| < \varepsilon/2$ for some $i \in \{1, \ldots, m\}$. For each $i \in \{1, \ldots, m\}$, let $y_i^* \in S_{X^*}$ be such that $\langle y_i, y_i^* \rangle = 1$; as $\dim X = \infty$, there exists $x \in S_X$ such that $\langle x, y_i^* \rangle = 0$ for all $i \in \{1, \ldots, m\}$. We assert that this x has the desired property. To verify this, let $y \in B \cap S_X$, let $i \in \{1, \ldots, m\}$ be such that $\|y - y_i\| < \varepsilon/2$ and let λ be a scalar. Then

$$\|y + \lambda x\| \geq \|y_i + \lambda x\| - \varepsilon/2 \geq \langle y_i + \lambda x, y_i^* \rangle - \varepsilon/2 = 1 - \varepsilon/2 \geq 1/(1 + \varepsilon).$$

Thus given $y \in B \backslash \{0\}$ and a scalar λ,

$$\left\| \frac{y}{\|y\|} + \frac{\lambda}{\|y\|} x \right\| \geq 1/(1 + \varepsilon),$$

from which the assertion follows. $\qquad \square$

Theorem 1.2.6 (Mazur). *Every infinite-dimensional Banach space X contains a basic sequence.*

Proof. Let $\varepsilon > 0$ and let $(\varepsilon_n)_{n\in\mathbb{N}}$ be a sequence of positive numbers such that $\prod_{n=1}^{\infty}(1+\varepsilon_n) \leq 1+\varepsilon$; let $x_1 \in S_X$. Using Proposition 1.2.5 construct inductively a sequence $(x_n)_{n\in\mathbb{N}}$ in S_X such that for all $n \in \mathbb{N}$ and all scalars λ,

$$\|y\| \leq (1+\varepsilon_n)\|y+\lambda x_{n+1}\| \text{ for all } y \in \mathrm{sp}\{x_1,\ldots,x_n\}.$$

An inductive argument shows that whenever $m, n \in \mathbb{N}$ with $n > m$, and $a_1,\ldots,a_m$ are scalars,

$$\left\|\sum_{i=1}^{n} a_i x_i\right\| \leq (1+\varepsilon_n)\cdots(1+\varepsilon_{m-1})\left\|\sum_{i=1}^{m} a_i x_i\right\| \leq (1+\varepsilon)\left\|\sum_{i=1}^{m} a_i x_i\right\|.$$

By Proposition 1.2.4, $(x_n)_{n\in\mathbb{N}}$ is a basic sequence. Its basis constant is bounded above by $1+\varepsilon$ since the corresponding canonical projections P_n satisfy $\|P_n\| \leq \prod_{i=n}^{\infty}(1+\varepsilon_i) \leq 1+\varepsilon$. $\square$

To establish the next main result, the Bessaga–Pelczyński selection principle, the following variant of Lemma 1.2.5 will be useful.

Lemma 1.2.7. *Let $(x_n)_{n\in\mathbb{N}}$ be a sequence in a Banach space X such that*

$$\liminf_{n\to\infty}\|x_n\| > 0 \quad \text{and} \quad x_n \rightharpoonup 0.$$

Then given any $\varepsilon \in (0,1)$, any $N \in \mathbb{N}$ and any finite-dimensional subspace B of X, there exists $n \geq N$ such that $\|y+\lambda x_n\| \geq (1-\varepsilon)\|y\|$ for all $y \in B$ and all scalars λ.

Proof. Without loss of generality we may suppose that $\|x_n\| = 1$ for all n. Let $\varepsilon \in (0,1)$, $N \in \mathbb{N}$ and let $\{y_1,\ldots,y_m\}$ be a (finite) $\varepsilon/3$-net of $B \cap S_X$. To each $i \in \{1,\ldots,m\}$ there corresponds $y_i^* \in S_{X^*}$ such that $\langle y_i, y_i^*\rangle_X \geq 1-\varepsilon/3$; since $x_n \rightharpoonup 0$, there exists $n \geq N$ such that $|\langle x_n, y_i^*\rangle_X| \leq \varepsilon/6$ if $i \in \{1,\ldots,m\}$. Let $y \in B \cap S_X$. If $|\lambda| \geq 2$, then

$$\|y+\lambda x_n\| \geq |\lambda|\,\|x_n\| - \|y\| \geq 2-1 \geq (1-\varepsilon)\|y\|;$$

while if $|\lambda| < 2$, let $i \in \{1,\ldots,m\}$ be such that $\|y-y_i\| < \varepsilon/3$ and observe that

$$\|y+\lambda x_n\| \geq |\langle y+\lambda x_n, y_i^*\rangle_X| \geq |\langle y_i, y_i^*\rangle_X| - |\langle \lambda x_n, y_i^*\rangle_X| - |\langle y-y_i, y_i^*\rangle_X| > 1-\varepsilon,$$

so that in either case the lemma holds. $\square$

Theorem 1.2.8 (The Bessaga–Pelczyński selection principle). *Let $(x_n)_{n\in\mathbb{N}}$ be a sequence in a Banach space X such that $\liminf_{n\to\infty}\|x_n\| > 0$ and $x_n \rightharpoonup 0$. Then $(x_n)_{n\in\mathbb{N}}$ contains a subsequence that is a basic sequence.*

Proof. Let $(\varepsilon_n)_{n\in\mathbb{N}}$ be a sequence of positive numbers such that

$$\delta := \prod_{n=1}^{\infty}(1-\varepsilon_n) > 0.$$

The desired subsequence of $(x_n)_{n\in\mathbb{N}}$ is defined by induction. Let $n_1 = 1$, $N = n_1$, $B = \mathrm{sp}\{x_1\}$ and $\varepsilon = \varepsilon_1$. By Lemma 1.2.7, there exists n_2 such that

$$\|\lambda_1 x_{n_1} + \lambda_2 x_{n_2}\| \geq (1-\varepsilon_1)\|\lambda_1 x_{n_1}\|$$

for all scalars λ_1, λ_2. Suppose that natural numbers n_i $(1 \leq i \leq k)$, with $n_1 < n_2 < \cdots < n_k$, have been chosen so that

$$\|\lambda_1 x_{n_1} + \cdots + \lambda_k x_{n_k}\| \geq (1-\varepsilon_{k-1})\|\lambda_1 x_{n_1} + \cdots + \lambda_{k-1} x_{n_{k-1}}\|$$

for all scalars $\lambda_1, \ldots, \lambda_k$. Let B be the linear span of $x_{n_1}, \ldots, x_{n_k}$, and let $\varepsilon = \varepsilon_k$, $N = n_k$. By Lemma 1.2.7, there exists $n_{k+1} > n_k$ such that

$$\|\lambda_1 x_{n_1} + \cdots + \lambda_{k+1} x_{n_{k+1}}\| \geq (1-\varepsilon_k)\|\lambda_1 x_{n_1} + \cdots + \lambda_k x_{n_k}\|.$$

We claim that the subsequence $(x_{n_k})_{k\in\mathbb{N}}$ thus defined is basic. Note that for any $k, m \in \mathbb{N}$ and any scalars $\lambda_1, \ldots, \lambda_{k+m}$,

$$\left\|\sum_{i=1}^{k+m}\lambda_i x_{n_i}\right\| \geq (1-\varepsilon_{k+m-1})\left\|\sum_{i=1}^{k+m-1}\lambda_i x_{n_i}\right\|$$

$$\geq (1-\varepsilon_{k+m-1})\cdots(1-\varepsilon_k)\left\|\sum_{i=1}^{k}\lambda_i x_{n_i}\right\| \geq \delta\left\|\sum_{i=1}^{k}\lambda_i x_{n_i}\right\|.$$

Now use Proposition 1.2.4. □

A pair of families $(x_n)_{n\in\mathbb{N}}$, $(x_n^*)_{n\in\mathbb{N}}$ in a Banach space X and X^* respectively such that $\langle x_i, x_j^*\rangle = \delta_{i,j}$ $(i, j \in \mathbb{N})$ is called a **biorthogonal system**, and the x_i^* are called **biorthogonal functionals** (with respect to (x_n)). For example, when $(x_n)_{n\in\mathbb{N}}$ is a basis of X, so that each $x \in X$ has a unique representation in the form $x = \sum_{n=1}^{\infty} a_n x_n \in X$, for each $n \in \mathbb{N}$ we define a functional x_n^* by $\langle x, x_n^*\rangle = a_n$. In terms of the canonical projections associated with this basis we have

$$\|P_n x - P_{n-1}x\| = \|\langle x, x_n^*\rangle x_n\| = |\langle x, x_n^*\rangle|\, \|x_n\|,$$

and so

$$\|x_n^*\| = \sup_{x\in B_X}|\langle x, x_n^*\rangle| = \|x_n\|^{-1}\sup_{x\in B_X}\|P_n x - P_{n-1}x\| \leq 2\|x_n\|^{-1}\sup_{n\in\mathbb{N}}\|P_n\| < \infty.$$

Hence $x_n^* \in X^*$ for all $n \in \mathbb{N}$ and

$$\|x_n\|_X\,\|x_n^*\|_{X^*} \leq 2\sup_n\|P_n\| = 2bc(x_n). \tag{1.2.2}$$

The x_n^* are biorthogonal functionals (with respect to the basis (x_n)) and are uniquely determined by the conditions $\langle x_m, x_n^*\rangle = \delta_{m,n}$.

Proposition 1.2.9. *Let (x_n) be a basis of a Banach space X with associated canonical projections P_n and biorthogonal functionals x_n^*. Then:*

(i) *for every $n \in \mathbb{N}$ and each $f \in X^*$,*

$$P_n^* f = \sum_{i=1}^{n} \langle x_i, f \rangle\, x_i^*;$$

(ii) *for all $f \in X^*$, $(P_n^* f)$ converges to f in the weak* sense;*

(iii) *(x_n^*) is a Schauder basis of the closed linear span, $\overline{\mathrm{sp}}\{x_n^* : n \in \mathbb{N}\} := S$ of the x_n^*, with associated canonical projections P_n^*. In particular, $P_n^* f \to f$ for all $f \in S$.*

Proof. (i) Let $f \in X^*$, $n \in \mathbb{N}$ and $x = \sum_{i=1}^{\infty} \langle x, x_i^* \rangle\, x_i$. Then

$$\langle x, P_n^* f \rangle = \langle P_n x, f \rangle = \left\langle \sum_{i=1}^{n} \langle x, x_i^* \rangle\, x_i, f \right\rangle = \sum_{i=1}^{n} \langle x_i, f \rangle \langle x, x_i^* \rangle.$$

(ii) Since f is continuous,

$$\lim_{n \to \infty} \langle x, P_n^* f \rangle = \lim_{n \to \infty} \sum_{i=1}^{n} \langle x_i, f \rangle \langle x, x_i^* \rangle = \left\langle \lim_{n \to \infty} \sum_{i=1}^{n} \langle x, x_i^* \rangle\, x_i, f \right\rangle = \langle x, f \rangle.$$

(iii) Routine arguments show that $P_n^* P_m^* = P_{\min(m,n)}^*$. If f belongs to the linear span of the x_n^*, then $P_n^* f = f$ for large enough n, so that $\|P_n^* f - f\| \to 0$. Since $\|P_n^*\| = \|P_n\|$, the P_n^* are uniformly bounded, and the desired result follows from the converse part of Lemma 1.2.1 together with Lemma 1.2.2. $\square$

Definition 1.2.10. Let (x_n) be a basis of a Banach space X with associated biorthogonal functionals x_n^*. It is called **shrinking** if $\overline{\mathrm{sp}}\{x_n^* : n \in \mathbb{N}\} = X^*$, and is said to be **boundedly complete** if $\sum_{i=1}^{\infty} a_i x_i$ converges whenever the scalars a_i are such that $\sup_n \|\sum_{i=1}^{n} a_i x_i\| < \infty$.

Proposition 1.2.11. *Let (x_n) be a basis of a Banach space X with associated canonical projections P_n and biorthogonal functionals x_n^*. Then the following statements are equivalent:*

(i) *(x_n) is shrinking;*

(ii) *(x_n^*) is a Schauder basis of X^*;*

(iii) *$\lim_{n \to \infty} \|f\!\restriction_{\overline{\mathrm{sp}}\{x_i : i > n\}}\| = 0$ for all $f \in X^*$.*

Proof. (i)$\Longrightarrow$(ii) follows from Proposition 1.2.9 (iii).
(ii)$\Longrightarrow$(i). Since $P_n^* f \to f$ for all $f \in X^*$, it follows that $X^* = \overline{\mathrm{sp}}\{x_n^* : n \in \mathbb{N}\}$.
(i)$\Longleftrightarrow$(iii). Let $f \in X^*$. If P is a bounded linear projection of X onto $P(X)$, then

$$\sup\{|\langle Px, f \rangle| : x \in B_X\} = \sup\{|\langle x, P^* f \rangle| : x \in B_X\} = \|P^* f\|$$

and

$$B_{P(X)} \subset P(B_X) \subset (\|P\|\, B_X \cap P(X)) \subset \|P\|\, B_{P(X)}.$$

With $id : X \to X$ as the identity map, it follows that for all n,

$$\begin{aligned}
\|f \upharpoonright_{(id-P_n)(X)}\| &= \sup\left\{|\langle x, f\rangle| : x \in B_{(id-P_n)(X)}\right\} \\
&\le \sup\left\{|\langle x, f\rangle| : x \in (id - P_n)(B_X)\right\} \\
&\le \sup\left\{|\langle x, f\rangle| : x \in (\|P_n\| + 1)\, B_{(id-P_n)(X)}\right\}.
\end{aligned}$$

Thus

$$\|f \upharpoonright_{(id-P_n)(X)}\| \le \|f - P_n^* f\| \le (\|P_n\| + 1)\, \|f \upharpoonright_{(id-P_n)(X)}\|,$$

so that (x_n) is shrinking if and only if $\|f \upharpoonright_{(id-P_n)(X)}\| \to 0$ for every $f \in X^*$. $\square$

Proposition 1.2.12. *Let (x_n) be a basis of a Banach space X with associated canonical projections P_n and biorthogonal functionals x_n^*; suppose that (x_n) is shrinking and let S be the space of all scalar sequences (a_n) such that*

$$\|(a_i)\|_S := \sup_n \left\| \sum_{i=1}^n a_i x_i \right\| < \infty,$$

normed by $\|\cdot\|_S$. Then the map $T : X^{**} \to S$ given by $T(x^{**}) = (\langle x_i^*, x^{**}\rangle)$ is an isomorphism of X^{**} onto S; if (x_n) is monotone, T is an isometry.

Proof. That $(S, \|\cdot\|_S)$ is a normed linear space can be verified easily. Let $K = bc(x_n)$, $x \in X$, $x^* \in X^*$ and $x^{**} \in X^{**}$. Then, identifying X with its canonical image in X^{**},

$$P_n^*(x^*) = \sum_{i=1}^n \langle x_i, x^*\rangle x_i^*, \quad \langle x^*, P_n^{**} x^{**}\rangle = \sum_{i=1}^n \langle x_i^*, x^{**}\rangle \langle x_i, x^*\rangle,$$

so that $P_n^{**} x^{**} = \sum_{i=1}^n \langle x_i^*, x^{**}\rangle x_i$. Hence

$$\|T(x^{**})\|_S = \sup_n \left\| \sum_{i=1}^n \langle x_i^*, x^{**}\rangle x_i \right\| = \sup_n \|P_n^{**} x^{**}\| \le K\, \|x^{**}\|,$$

which shows that T is bounded.

Let $(a_i) \in S$. Since (x_n) is shrinking, X^* is separable. As $\left(\sum_{i=1}^n a_i x_i\right)_{n \in \mathbb{N}}$ is bounded in X^{**} (again identifying X with its canonical image in X^{**}), there is a weak* cluster point x^{**} of $\left(\sum_{i=1}^n a_i x_i\right)_{n \in \mathbb{N}}$ which satisfies $\langle x_i^*, x^{**}\rangle = a_i$ for each i. Also,

$$\|x^{**}\| \le \limsup_{n \to \infty} \left\| \sum_{i=1}^n a_i x_i \right\| \le \|(a_i)\|_S.$$

It follows that $T(x^{**}) = (a_i)$ and $\|Tx^{**}\| \ge \|x^{**}\|$: the proof is complete. $\square$

Theorem 1.2.13. *Let (x_n) be a basis of a Banach space X. Then X is reflexive if and only if (x_n) is both shrinking and boundedly complete.*

Proof. First suppose that X is reflexive. By Proposition 1.2.9 (ii), for every $f \in X^*$, $(P_n^* f)$ converges in the weak* sense to f in X^*, and hence also in the weak sense as X is reflexive. Hence X^* is the weak closure of the span of the x_n^*, which is just the closure of the span of the x_n^*. Thus (x_n) is shrinking. By Proposition 1.2.12, X^{**} is isomorphic to S: under this isomorphism, $X \subset X^{**}$ corresponds to $S_1 := \{(a_i) : \sum_{i=1}^{\infty} a_i x_i$ converges$\}$. Since X is reflexive, $S_1 = S$ and so (x_n) is boundedly complete.

Conversely, if (x_n) is shrinking and boundedly complete, the above identification holds and $S_1 = S$. Hence $X = X^{**}$ and X is reflexive.   □

As an obvious consequence of Proposition 1.2.11 and the last theorem we have

Corollary 1.2.14. *Let (x_n) be a basis of a reflexive Banach space X with associated biorthogonal functionals x_n^*. Then (x_n^*) is a Schauder basis of X^*.*

Companion to the notion of a basis is that of a **weak basis**: a sequence $(x_n)_{n \in \mathbb{N}}$ in a normed linear space X is said to be a weak basis of X if, given any $x \in X$, there is a unique sequence $(a_n)_{n \in \mathbb{N}}$ of scalars such that $\sum_{n=1}^{N} a_n x_n$ converges weakly to x as $N \to \infty$. The next result shows that this seemingly new concept coincides with the old.

Theorem 1.2.15. *Let (x_n) be a weak basis of a Banach space X. Then (x_n) is a basis of X.*

Proof. Step 1. We show that the coefficients $a_n(x)$ in the weak representation $\sum_{n=1}^{\infty} a_n(x) x_n$ of x depend continuously on x, so that each $a_n \in X^*$. To do this, let L be the linear space of all scalar sequences $\alpha = (\alpha_n)_{n \in \mathbb{N}}$ such that $\sum_{n=1}^{\infty} \alpha_n x_n$ exists in the weak sense. By the Banach–Steinhaus theorem,

$$\|\alpha\|_w := \sup_{n \in \mathbb{N}} \left\| \sum_{k=1}^{n} \alpha_k x_k \right\| < \infty;$$

routine verification shows that $\|\cdot\|_w$ is a norm on L. Define $T : L \to X$ by

$$T(\alpha) = (\text{weak}) \sum_{n=1}^{\infty} \alpha_n x_n.$$

As (x_n) is a weak basis, T is bijective; since

$$\|T(\alpha)\| = \left\| (\text{weak}) \sum_{n=1}^{\infty} \alpha_n x_n \right\| \leq \|\alpha\|_w,$$

T is also continuous. Assume for the moment that $(L, \|\cdot\|_w)$ is complete. Then by Banach's isomorphism theorem, T^{-1} is continuous, so that for each $n \in \mathbb{N}$, with $x = T(\alpha)$,

$$|\alpha_n| \, \|x_n\| = \left\| \sum_{k=1}^{n} \alpha_k x_k - \sum_{k=1}^{n-1} \alpha_k x_k \right\| \leq 2 \|\alpha\|_w \leq 2 \|T^{-1}\| \, \|x\|.$$

Hence $a_n \in X^*$.

It remains to show that $(L, \|\cdot\|_w)$ is complete. Let $\left(\alpha^{(k)}\right)_{k\in\mathbb{N}}$ be a Cauchy sequence in L, $\alpha^{(k)} = \left(\alpha_n^{(k)}\right)_{n\in\mathbb{N}}$. For each $n \in \mathbb{N}$,

$$\left|\alpha_n^{(k)} - \alpha_n^{(l)}\right| \|x_n\| \le 2 \left\|\alpha^{(k)} - \alpha^{(l)}\right\|_w,$$

and so $\left(\alpha_n^{(k)}\right)_{k\in\mathbb{N}}$ is a Cauchy sequence of scalars: let α_n be its limit and set $\alpha = (\alpha_n)_{n\in\mathbb{N}}$. We claim that $\alpha \in L$ and that $\left\|\alpha^{(k)} - \alpha\right\|_w \to 0$. To prove this, let $\varepsilon > 0$. There exists $N = N(\varepsilon)$ such that $\left\|\alpha^{(k)} - \alpha^{(N)}\right\|_w < \varepsilon$ if $k \ge N$. Thus for all $n \in \mathbb{N}$ and all $k \ge N$,

$$\left\|\sum_{i=1}^n \left(\alpha_i^k - \alpha_i^N\right) x_i\right\| < \varepsilon,$$

so that for all $n \in \mathbb{N}$,

$$\left\|\sum_{i=1}^n \left(\alpha_i - \alpha_i^N\right) x_i\right\| \le \varepsilon. \tag{1.2.3}$$

Since T is continuous, $\left(T(\alpha^{(k)})\right)$ is a Cauchy sequence in X; let x be its limit. Thus for some $M = M(\varepsilon) \ge N$,

$$\left\|x - T(\alpha^{(M)})\right\| < \varepsilon. \tag{1.2.4}$$

Let $x^* \in X^*$, $\|x^*\| \le 1$. As (x_n) is a weak basis, there exists $R \in \mathbb{N}$ such that if $n \ge R$, then

$$\left|\left\langle \sum_{i=1}^n \alpha_i^{(M)} x_i - T(\alpha^{(M)}), x^* \right\rangle\right| < \varepsilon. \tag{1.2.5}$$

Use of equations $(1.2.3)$–$(1.2.5)$ now gives, for all $n \ge R$,

$$\left|\left\langle \sum_{i=1}^n \alpha_i x_i - x, x^* \right\rangle\right| \le \left|\left\langle \sum_{i=1}^n \left(\alpha_i - \alpha_i^{(M)}\right) x_i, x^* \right\rangle\right|$$
$$+ \left|\left\langle \sum_{i=1}^n \alpha_i^{(M)} x_i - T(\alpha^{(M)}), x^* \right\rangle\right| + \left|\left\langle T(\alpha^{(M)}) - x, x^* \right\rangle\right|$$
$$\le \left\|\sum_{i=1}^n \left(\alpha_i - \alpha_i^{(M)}\right) x_i\right\| + \left|\left\langle \sum_{i=1}^n \alpha_i^{(M)} x_i - T(\alpha^{(M)}), x^* \right\rangle\right|$$
$$+ \left|\left\langle T(\alpha^{(M)}) - x, x^* \right\rangle\right|$$
$$< 3\varepsilon.$$

Thus $x = (\text{weak})\sum_{i=1}^\infty \alpha_i x_i = T(\alpha)$ and $\alpha \in L$: from $(1.2.3)$ it follows that $\left\|\alpha - \alpha^{(k)}\right\|_w \to 0$. Hence $(L, \|\cdot\|_w)$ is complete and Step 1 is accomplished.

Step 2. We prove that (x_n) is a basis of X. Define $P_n : X \to X$ by $P_n(x) = \sum_{k=1}^{n} a_k x_k$, where $x \in X$ is the weak limit of $\sum_{n=1}^{N} a_n x_n$ as $N \to \infty$. For all $x \in X$ the sequence $(P_n x)_{n \in \mathbb{N}}$ is weakly convergent and hence bounded: by uniform boundedness, $K := \sup_n \|P_n\| < \infty$. Let

$$E = \{x \in X : (P_n x)_{n \in \mathbb{N}} \text{ converges in } X\}.$$

We claim that E is a closed linear subspace of X. Assume this for the moment. Since $\|P_m x_n - x_n\| = 0$ when $m \geq n$, we see that every x_n belongs to E; since E is closed and linear it is weakly closed, and hence coincides with X. Thus (x_n) is a basis of X.

It remains to establish the claimed property of E. That it is linear is clear. For the closedness, let $(y_i)_{i \in \mathbb{N}}$ be a sequence of points of E with limit y, and suppose that $z_i = \lim_{n \to \infty} P_n y_i$ $(i \in \mathbb{N})$. Since

$$\|z_i - z_j\| \leq K \|y_i - y_j\|,$$

(z_i) is a Cauchy sequence in X; let z be its limit. Then

$$\|P_n y - z\| \leq \|P_n(y - y_i)\| + \|P_n y_i - z_i\| + \|z_i - z\|$$
$$\leq K \|y - y_i\| + \|P_n y_i - z_i\| + \|z_i - z\|,$$

from which a routine argument gives $P_n y \to z$, so that $y \in E$. The proof of the Theorem is complete. $\qquad\square$

The task of showing that a given sequence of elements of a Banach space X is a basis is simplified if another basis is known explicitly, for then comparisons between the two sequences may be made: for example, perturbation techniques can be used (see [58], V.2.5, for instance, when X is a Hilbert space). The following elementary result is useful in this connection.

Proposition 1.2.16. *Let (x_n) be a basis of a Banach space X with associated biorthogonal functionals x_n^*, let T be a linear homeomorphism of X onto a Banach space Y and for each $n \in \mathbb{N}$ put $y_n = T x_n$. Then (y_n) is a basis of Y.*

Proof. Let $y \in Y$. Then $y = T x$ for some unique $x \in X$. Hence

$$\left\| y - \sum_{k=1}^{n} \langle x, x_k^* \rangle y_k \right\|_Y = \left\| T \left(x - \sum_{k=1}^{n} \langle x, x_k^* \rangle x_k \right) \right\|_Y$$
$$\leq \|T\| \left\| x - \sum_{k=1}^{n} \langle x, x_k^* \rangle x_k \right\|_X \to 0$$

as $n \to \infty$. Thus each $y \in Y$ is representable in the form $y = \sum_{k=1}^{\infty} \langle x, x_k^* \rangle y_k$. To show that this representation is unique, suppose that $y = \sum_{k=1}^{\infty} a_k y_k$. Then

$$\left\| \sum_{k=1}^{n} (a_k - \langle x, x_k^* \rangle) x_k \right\|_X = \left\| T^{-1} \left(\sum_{k=1}^{n} (a_k - \langle x, x_k^* \rangle) y_k \right) \right\|_X$$

$$\leq \|T^{-1}\| \left\| \sum_{k=1}^{n} (a_k - \langle x, x_k^* \rangle) y_k \right\|_Y \to 0$$

as $n \to \infty$. Since (x_n) is a basis, $a_k = \langle x, x_k^* \rangle$ for all k. $\qquad \square$

We next turn to more detailed information concerning the biorthogonal functionals corresponding to a basis. In this we shall need the notion of a **seminormalised basis** (x_n), by which we mean that $0 < \inf_n \|x_n\| \leq \sup_n \|x_n\| < \infty$.

Proposition 1.2.17. *Let (x_n) be a seminormalised basis of a uniformly convex Banach space X. Then there exist $p \in (1, \infty)$ and $K > 0$ such that for every $x = \sum_{i=1}^{\infty} a_i x_i \in X$,*

$$\|x\| \leq K \left(\sum_{i=1}^{\infty} |a_i|^p \right)^{1/p}.$$

Proof. Without loss of generality we may assume that the basis is normalised in the sense that $\|x_n\| = 1$ for all n. Let A be the basis constant for the canonical projections P_n, take $\varepsilon \in (0, 1/A)$, let $\lambda = 2(1 - \delta(\varepsilon))$, where δ is the modulus of convexity of X, and let $p \in (1, \log_\lambda 2)$. We claim that there exists $\alpha \in (0, 1)$ such that $\|x + ty\|^p < 1 + t^p$ whenever $|t - 1| < \alpha$ and $x, y \in S_X$ are representable in terms of the basis elements by finite sums over disjoint sets of indices, every element of one index set being less than every element of the second. We suppose that l is the maximum index for x and that it is less than the minimum index for y. Then

$$A \|x - y\| \geq \|P_l(x - y)\| = \|x\| = 1,$$

so that $\|x - y\| \geq 1/A > \varepsilon$, and hence $\|x + y\| \leq \lambda$. As $\lambda^p < 2$, it follows that $\|x + y\|^p < 2 = 1 + 1^p$. Since $t \longmapsto \|x + ty\|^p$ is uniformly continuous at $t = 1$ with respect to $x, y \in S_X$, there exists $\alpha \in (0, 1)$ such that $\|x + ty\|^p < 1 + t^p$ whenever $|t - 1| < \alpha$ and $x, y \in S_X$, as claimed.

Put $K = 2/\alpha$. It is enough to show that $\|z_n\| \leq K \left(\sum_{i=1}^{n} |\alpha_i|^p \right)^{1/p}$ for every finitely expressible element $z_n = \sum_{i=1}^{n} \alpha_i x_i$. We use induction. The result is plainly true for $n = 1$. Assume it holds for some n and let $z_{n+1} = \sum_{i=1}^{n+1} \alpha_i x_i \neq 0$. Two cases can be distinguished:

(1) $|\alpha_i| \leq \|z_{n+1}\| / K$ for all $i \in \{1, 2, \ldots, n + 1\}$;
(2) $|\alpha_{i_0}| > \|z_{n+1}\| / K$ for some $i_0 \in \{1, 2, \ldots, n + 1\}$.

For case (1), put $z_0 = 0, z_s = \sum_{i=1}^{s} \alpha_i x_i$ for all $s \in \{1, 2, \ldots, n+1\}$, $y_s = \sum_{i=s+1}^{n+1} \alpha_i x_i$ for all $s \in \{0, 1, 2, \ldots, n\}$, $y_{n+1} = 0$. Then

$$\|z_0\| < \|y_0\|, \|z_{n+1}\| > \|y_{n+1}\|, |\|z_{i+1}\| - \|z_i\|| \le \|z_{n+1}\|/K$$

and

$$|\|y_{i+1}\| - \|y_i\|| \le \|z_{n+1}\|/K$$

for all $i \in \{1, 2, \ldots, n\}$.

We now make the following assertion: Let $\{\xi_i\}_{i=1}^{n}$ and $\{\eta_i\}_{i=1}^{n}$ be sequences of real numbers such that for some $\varepsilon > 0$, $\xi_1 < \eta_1, \xi_n > \eta_n$ and $|\xi_{i+1} - \xi_i| < \varepsilon, |\eta_{i+1} - \eta_i| < \varepsilon$ for all $i \in \{1, 2, \ldots, n-1\}$. Then there exists $i_0 \in \{1, 2, \ldots, n\}$ such that $|\xi_{i_0} - \eta_{i_0}| < \varepsilon$.

The claim is established by consideration of the first index j for which $\xi_j > \eta_j$. In view of this assertion, there exists $r \in \{1, 2, \ldots, n\}$ such that

$$|\|z_r\| - \|y_r\|| \le \|z_{n+1}\|/K.$$

By interchanging z_r and y_r if necessary we can suppose that $\|z_r\| > \|y_r\|$; and by homogeneity we may assume that $\|z_r\| = 1$. Since $\|z_r\| > \|y_r\|$ and $z_{n+1} = z_r + y_r$, we have $\|z_{n+1}\| \le 2$ so that $|1 - \|y_r\|| \le 2/K = \alpha$. Now put $t = \|y_r\|$ and $\widetilde{y}_r = y_r/\|y_r\|$. Application of the first part of the proof to $x = z_r$ and $y = \widetilde{y}_r$, noting that $|1 - t| \le \alpha$, shows that $\|z_r + t\widetilde{y}_r\|^p < 1 + t^p$. Hence

$$\|z_{n+1}\|^p = \|z_r + y_r\|^p = \|z_r + t\widetilde{y}_r\|^p < 1 + t^p = \|z_r\|^p + \|y_r\|^p.$$

By the inductive hypothesis,

$$\|z_r\|^p + \|y_r\|^p \le K^p \sum_{i=1}^{r} |\alpha_i|^p + K^p \sum_{i=r+1}^{n} |\alpha_i|^p.$$

Thus $\|z_{n+1}\|^p \le K^p \sum_{i=1}^{n+1} |\alpha_i|^p$ and the treatment of case (1) is complete. For case (2), assuming that $|\alpha_{i_0}| > \|z_{n+1}\|/K$, we have

$$\|z_{n+1}\| \le K |\alpha_{i_0}| \le K \left(\sum_{i=1}^{n+1} |\alpha_i|^p \right)^{1/p}.$$

The inductive step is therefore complete and the proposition follows.   □

Proposition 1.2.18. *Let (x_n) be a seminormalised basis of a uniformly smooth Banach space X. Then there exist $q \in (1, \infty)$ and $L > 0$ such that for every $x = \sum_{i=1}^{\infty} \alpha_i x_i \in X$,*

$$\|x\| \ge L \left(\sum_{i=1}^{\infty} |\alpha_i|^q \right)^{1/q}.$$

Proof. Let x_n^* be the biorthogonal functionals associated to the basis (x_n). Since X is uniformly smooth it is reflexive, and so by Corollary 1.2.14 and (1.2.2), the x_n^* form a seminormalised Schauder basis of X^*. As X^* is uniformly convex, by Proposition 1.2.17 there exist $p \in (1, \infty)$ and $K > 0$ such that for every $x^* = \sum_{i=1}^\infty \beta_i x_i^* \in X^*$,

$$\|x^*\| \leq K \left(\sum_{i=1}^\infty |\beta_i|^p \right)^{1/p}.$$

We claim that $L = 1/K$ and $q = p'$ have the properties asserted by the proposition. To verify this, let $x = \sum_{i=1}^\infty \alpha_i x_i \in X$, take any $n \in \mathbb{N}$ and let $z_n = \sum_{i=1}^n \alpha_i x_i$, $g_n = \sum_{i=1}^n \beta_i x_i^* \in X^*$. Then

$$|\langle z_n, g_n \rangle| / \|z_n\| \leq \|g_n\| \leq K \left(\sum_{i=1}^n |\beta_i|^p \right)^{1/p}.$$

Hence

$$\|z_n\| \geq \frac{|\langle z_n, g_n \rangle|}{K \left(\sum_{i=1}^n |\beta_i|^p \right)^{1/p}} = \frac{|\sum_{i=1}^n \alpha_i \beta_i|}{K \left(\sum_{i=1}^n |\beta_i|^p \right)^{1/p}}.$$

Now take $\beta_i = |\alpha_i|^{1/(p-1)} \operatorname{sgn}(\alpha_i)$ for $i = 1, 2, \ldots, n$. Then

$$\|z_n\| \geq \frac{\sum_{i=1}^n |\alpha_i|^q}{K \left(\sum_{i=1}^n |\alpha_i|^q \right)^{1/p}} = \left(\sum_{i=1}^n |\alpha_i|^q \right)^{1/q} / K.$$

As this holds for all $n \in \mathbb{N}$ the result follows. □

As an immediate consequence of the last two propositions we have

Theorem 1.2.19. *Let (x_n) be a seminormalised basis of a uniformly smooth, uniformly convex Banach space X. Then there exist $p, q \in (1, \infty)$ and $K > 0$ such that for every $x = \sum_{i=1}^\infty a_i x_i \in X$,*

$$K^{-1} \left(\sum_{i=1}^\infty |a_i|^q \right)^{1/q} \leq \|x\| \leq K \left(\sum_{i=1}^\infty |a_i|^p \right)^{1/p}.$$

When $p = q = 2$ this is just the condition satisfied by a Riesz basis in a Hilbert space.

To conclude this section we briefly discuss an approximation property that a Banach space may posses, and its relation to the existence of a basis.

Definition 1.2.20. A Banach space X is said to have the **approximation property** **(AP)** if, given any compact subset K of X and any $\varepsilon > 0$, there exists $T \in B(X)$ with finite rank such that $\|Tx - x\| < \varepsilon$ for all $x \in K$.

Simple equivalent forms of the AP are given in the next proposition.

Proposition 1.2.21. *Let X be a Banach space. Then the following are each equivalent to the statement that X has the AP.*

(i) *For any Banach space Y, every $T \in B(X,Y)$, every compact subset K of X and every $\varepsilon > 0$, there exists $F \in B(X,Y)$ with finite rank such that $\|Tx - Fx\| < \varepsilon$ for all $x \in K$.*

(ii) *For any Banach space Y, every $T \in B(Y,X)$, every compact subset K of Y and every $\varepsilon > 0$, there exists $F \in B(Y,X)$ with finite rank such that $\|Ty - Fy\| < \varepsilon$ for all $y \in K$.*

Proof. Suppose that X has the AP and that Y is a Banach space. Let $T \in B(X,Y)\setminus\{0\}$, K be a compact subset of X and $\varepsilon > 0$. Then there exists $R \in B(X)$ with finite rank such that $\|Rx - x\| < \varepsilon/\|T\|$ for all $x \in K$. Now take $F = T \circ R$: (i) holds.

Again suppose that X has the AP and that Y is a Banach space. Let $T \in B(Y,X)$, K be a compact subset of Y and $\varepsilon > 0$. Then $\overline{T(K)}$ is compact, and so there exists $R \in B(X)$ with finite rank such that $\|Rx - x\| < \varepsilon$ for all $x \in T(K)$. Now take $F = R \circ T$: (ii) holds.

Clearly each of (i) and (ii) implies that X has the AP. $\square$

Some preparatory lemmas will enable us to give further criteria for the AP.

Lemma 1.2.22. *Let X, Y be Banach spaces and let $T, T_n \in B(X,Y)$ $(n \in \mathbb{N})$ be such that $\lim_{n \to \infty} T_n x = Tx$ for all $x \in X$. Then for every compact $K \subset X$, $T_n x \to Tx$ uniformly on K.*

Proof. Suppose the result is false. Then there exist a compact set K, $\varepsilon > 0$, a subsequence of (T_n) (again denoted by (T_n)) and $x_n \in K$ $(n \in \mathbb{N})$ such that $\|T_n x_n - Tx_n\| \geq \varepsilon$ for all $n \in \mathbb{N}$. As K is compact, we may suppose that $x_n \to x \in K$. By uniform boundedness, $M := \max(\|T\|, \sup_n \|T_n\|) < \infty$. Then

$$\begin{aligned}
\|(T_n - T)(x_n)\| &\leq \|(T_n - T)(x)\| + \|(T_n - T)(x_n - x)\| \\
&\leq \|(T_n - T)(x)\| + 2M\|x_n - x\| \\
&\to 0,
\end{aligned}$$

and we have a contradiction. $\square$

Corollary 1.2.23. *Every Banach space X with a basis has the AP.*

Proof. Suppose that X has a basis (x_n), and let (P_n) be the family of associated projections. Then $P_n x \to x$ for all $x \in X$, so that Lemma 1.2.22 may be applied.
                                                                                          $\square$

Lemma 1.2.24. *Let K be a closed subset of a Banach space X. Then K is compact if and only if there is a sequence (x_n) in X such that $x_n \to 0$ and K is contained in the closed convex hull $\overline{\mathrm{conv}}\{x_n : n \in \mathbb{N}\} := C$ of the x_n.*

Proof. Suppose that K is compact and let $x_{i,1}$ $(i = 1, \ldots, n_1)$ be elements of X such that $2K \subset \cup_{i=1}^{n_1} B(x_{i,1}, 1/4)$. The set

$$K_2 := \cup_{i=1}^{n_1} \{(B(x_{i,1}, 1/4) \cap 2K) - x_{i,1}\}$$

is a compact subset of $B(0, 1/4)$. Now choose $x_{i,2}$ $(i = 1, \ldots, n_2)$ in $B(0, 1/2)$ so that $2K_2 \subset \cup_{i=1}^{n_2} B(x_{i,2}, 1/4^2)$ and let

$$K_3 := \cup_{i=1}^{n_2} \{(B(x_{i,2}, 1/4^2) \cap 2K_2) - x_{i,2}\}.$$

Points $x_{i,j}$ $(i = 1, \ldots, n_j; j \in \mathbb{N})$ are defined in the natural inductive way. Given any $x \in K$, there exists $i_1, 1 \leq i_1 \leq n_1$, such that $2x - x_{i_1,1} \in K_2$; there exists $i_2, 1 \leq i_2 \leq n_2$, such that $4x - 2x_{i_1,1} - x_{i_2,2} \in K_3$; and generally,

$$x - \left(x_{i_1,1}/2 + x_{i_2,2}/2^2 + \cdots + x_{i_k,k}/2^k\right) \in 2^{-k} K_{k+1}.$$

Thus $x \in \overline{\text{conv}} \{x_{i,j} : 1 \leq i \leq n_j, j \in \mathbb{N}\}$. Since $\|x_{i,j}\| \leq 2 \cdot 4^{-j+1}$ for $j > 1$ and $i \leq n_j$, the claimed implication follows. For the reverse implication, note that if $x_n \to 0$ then $\{\sum_{n=1}^{\infty} \lambda_n x_n : \lambda_n \geq 0, \sum_{n=1}^{\infty} \lambda_n \leq 1\}$ is compact and coincides with C. $\qquad\square$

Theorem 1.2.25. *A Banach space X has the AP if and only if for every Banach space Y, every compact $T \in B(Y, X)$ and every $\varepsilon > 0$, there is a finite rank operator $F \in B(Y, X)$ with $\|T - F\| < \varepsilon$.*

Proof. Suppose that X has the AP and let $T \in B(Y, X)$ be compact. Then $K := TB_Y(0, 1)$ is compact: by the AP, given $\varepsilon > 0$, there exists a finite rank operator $T_1 \in B(X)$ such that $\|T_1 x - x\| < \varepsilon$ if $x \in K$. Then $\|T_1 T - T\| < \varepsilon$.

For the converse, let K be a compact subset of X and $\varepsilon > 0$. In view of Lemma 1.2.24, we may suppose that $K = \overline{\text{conv}}\{x_n : n \in \mathbb{N}\}$, where $\|x_n\| \downarrow 0$ and $\|x_1\| \leq 1$. Put $U = \overline{\text{conv}} \{\pm x_n / \|x_n\|^{1/2} : n \in \mathbb{N}\}$: U is a compact convex set, symmetric about the origin. Let Y be the linear span of U in X, so that $Y = \cup_{n=1}^{\infty} nU$, and define a norm on Y by

$$\|y\|_1 = \inf \{\lambda > 0 : y/\lambda \in U\},$$

with respect to which U is the unit ball. It can be checked that $(Y, \|\cdot\|_1)$ is a Banach space. The identity map from Y to X is compact and so, by hypothesis, there exist $y_i^* \in Y^*$ and $u_i \in X$ $(i = 1, \ldots, m)$ such that

$$\left\| \sum_{i=1}^{m} \langle x, y_i^* \rangle_Y u_i - x \right\| < \varepsilon/2$$

for every $x \in U$ and hence every $x \in K$. Note that the y_i^* are elements of Y^* and need not be restrictions to Y^* of elements of X^*. To finish the proof it is enough to establish the following claim: given $y^* \in Y^*$ and $\delta > 0$ (we take $\delta = $

$\varepsilon/\{2m \max_i \|u_i\|\}$), there exists $x^* \in X^*$ such that $|\langle x, y^* \rangle_Y - \langle x, x^* \rangle_X| < \delta$ if $x \in K$, that is, $|\langle x_n, y^* \rangle_Y - \langle x_n, x^* \rangle_X| < \delta$ for every n. Since $x_n/\|x_n\|^{1/2} \in U$, we have $\|x_n\|_1 \le \|x_n\|^{1/2} \to 0$: thus $|\langle x_n, y^* \rangle_Y| < \delta/2$ for $n > N$, say. Put

$$K_0 = 2\delta^{-1}\overline{\mathrm{conv}}\,\{\pm x_n : n \ge N + 1\}$$

(the closures in $\|\cdot\|$ and $\|\cdot\|_1$ are the same) and

$$F = \{x \in \mathrm{sp}\{x_n : n = 1, \dots, N\} : \langle x, y^* \rangle_Y = 1\}.$$

Then with respect to $\|\cdot\|$, the sets F and K_0 are respectively closed and compact; also $K_0 \cap F = \emptyset$. By the geometric form of the Hahn–Banach theorem there is a $\|\cdot\|$-closed hyperplane $H = \{x : \langle x, x^* \rangle_X = 1\}$ in X such that $F \subset H$ and $H \cap K_0 = \emptyset$. Then $\langle x_n, x^* \rangle_X = \langle x_n, y^* \rangle_Y$ for $n \le N$ and $|\langle x_n, x^* \rangle_X| < \delta/2$ for $n > N$. Thus $|\langle x, y^* \rangle_Y - \langle x, x^* \rangle_X| < \delta$ for every n, as claimed.   □

If X is a Banach space such that X^* has the AP, then so does X; the converse is false. If X and Y are reflexive Banach spaces, one of which has the AP, then if $B(X, Y)$ is reflexive, every $T \in B(X, Y)$ is compact. For proofs of these assertions and further details of the AP and related notions we refer to [62], [74] and [80].

To conclude this section we remark that in a famous paper Enflo [45] showed that not every separable Banach space has the AP. Following the account given in [62] we indicate briefly how it can be proved that even such a familiar space as l_p has a subspace that does not have this property and consequently does not have a basis. The argument relies on the next two theorems.

Theorem 1.2.26. *The following two statements about a Banach space X are equivalent:*

(i) *X has the AP.*

(ii) *If (x_n) and (x_n^*) are sequences in X and X^* respectively such that*

$$\sum_{n=1}^{\infty} \|x_n\|\,\|x_n^*\| < \infty \quad and \quad \sum_{n=1}^{\infty} \langle x, x_n^* \rangle x_n = 0 \ for \ all \ x \in X,$$

then

$$\sum_{n=1}^{\infty} \langle x_n, x_n^* \rangle = 0.$$

Proof. Endow $B(X)$ with the topology τ of uniform convergence on compact sets: this is the locally convex topology generated by the seminorms $\|T\|_K := \sup\{\|Tx\| : x \in K\}$, where K is any compact subset of X. Then (see [62] I, Proposition 1.e.3) the dual of $(B(X), \tau)$ is the set of all functionals ϕ of the form

$$\phi(T) = \sum_{j=1}^{\infty} \langle Tx_i, x_i^* \rangle, \ (x_i) \subset X, (x_i^*) \subset X^*, \sum_{j=1}^{\infty} \|x_i\|\,\|x_i^*\| < \infty.$$

To say that X has the AP means that the identity map on X is in the τ-closure of the space of finite rank operators in $B(X)$. This is the case if and only if every τ-continuous linear functional on $B(X)$ which vanishes on operators of rank 1 also vanishes on the identity operator. In view of the characterisation of the dual of $(B(X), \tau)$ given above, this is what (ii) means. $\qquad\square$

Theorem 1.2.27. *There is an infinite matrix* $A = (a_{i,j})_{i,j\in\mathbb{N}}$ *with scalar entries such that:*

(a) *for any* $i \in \mathbb{N}$, $a_{i,j} = 0$ *except for* $j \in F(i)$, *where* $F(i)$ *is finite and non-empty;*

(b) *for every* $r \in (2/3, \infty)$, $\sum_{i=1}^{\infty} (\max_j |a_{i,j}|)^r < \infty$;

(c) $A^2 = 0$ *and* trace $A := \sum_{i=1}^{\infty} a_{i,i} \neq 0$.

The proof is technically quite complicated, as might be expected: we refer to [62] I, Theorem 2.d.3 for details.

Given these theorems, it is relatively straightforward to establish the following

Theorem 1.2.28. *Let* $p \in (2, \infty)$. *Then* l_p *has a subspace that does not have the AP.*

Proof. Let A be the matrix of Theorem 1.2.27; put $\lambda_i = \max_j |a_{i,j}|$ and $b_{i,j} = (\lambda_j/\lambda_i)^{1/(p+1)} a_{i,j}$ $(i, j \in \mathbb{N})$. The matrix $B := (b_{i,j})_{i,j\in\mathbb{N}}$ satisfies $B^2 = 0$ since the (i, j)th element of B^2 is

$$\sum_{k=1}^{\infty} b_{i,k} b_{k,j} = (\lambda_j/\lambda_i)^{1/(p+1)} \sum_{k=1}^{\infty} a_{i,k} a_{k,j},$$

which is 0 as $A^2 = 0$. Moreover, trace $B =$ trace $A \neq 0$. Let $y_i = (b_{i,1}, b_{i,2}, \ldots)$ and note that $(p+1)/p = 1 + 1/p < 3/2$, so that $p/(p+1) > 2/3$ and hence, by property (b) of Theorem 1.2.27, $\sum_{j=1}^{\infty} \lambda_j^{p/(p+1)} < \infty$. Thus

$$\|y_i\|_p = \left(\sum_{j=1}^{\infty} |b_{i,j}|^p \right)^{1/p} \leq \lambda_i^{p/(p+1)} \left(\sum_{j=1}^{\infty} \lambda_j^{p/(p+1)} \right)^{1/p} \leq \lambda_i^{p/(p+1)} L,$$

for some constant L independent of i, and so, invoking (b) again,

$$\sum_{i=1}^{\infty} \|y_i\|_p < \infty.$$

Now let $(e_i)_{i\in\mathbb{N}}$ be the standard basis of $l_{p'}$ and let X be the closed linear span (in l_p) of the y_i $(i \in \mathbb{N})$. We claim that for every $x \in X$,

$$\sum_{i=1}^{\infty} y_i \langle x, e_i \rangle = 0.$$

To verify this, note that if $x = y_k$, then

$$\sum_{i=1}^{\infty} y_i \langle y_k, e_i \rangle = \sum_{i=1}^{\infty} b_{k,i} y_i = \sum_{i=1}^{\infty} b_{k,i}(b_{i,1}, b_{i,2}, \ldots) = 0$$

since $B^2 = 0$. Linearity now gives the desired result for every element of the span of the y_i. Given any $\varepsilon > 0$ and any $x \in X$, there exists $x_n \in \mathrm{span}\,(y_i)$ such that $\|x - x_n\|_p < \varepsilon$. Since

$$\left\| \sum_{i=1}^{\infty} y_i \langle x, e_i \rangle \right\|_p = \left\| \sum_{i=1}^{\infty} y_i \langle x, e_i \rangle - \sum_{i=1}^{\infty} y_i \langle x_n, e_i \rangle \right\|_p$$

$$= \left\| \sum_{i=1}^{\infty} y_i \langle x - x_n, e_i \rangle \right\|_p \leq \varepsilon \sum_{i=1}^{\infty} \|y_i\|_p,$$

the claim follows. Moreover,

$$\sum_{i=1}^{\infty} \|y_i\|_p \|e_i\|_{p'} = \sum_{i=1}^{\infty} \|y_i\|_p < \infty$$

and

$$\sum_{i=1}^{\infty} \langle y_i, e_i \rangle = \sum_{i=1}^{\infty} b_{i,i} = \mathrm{trace}\,B \neq 0.$$

Thus by Theorem 1.2.26, X does not have a basis.   □

In [62] II, Theorem 1.g.4, this result is complemented by showing that when $1 \leq p < 2$, there is a subspace of l_p that does not have the AP.

1.3   The p-trigonometric functions

Here we introduce generalisations of the classical trigonometric functions that prove to be of considerable importance in the theory of the p-Laplacian on the line. Throughout this section we shall suppose that $1 < p < \infty$ and $p' = p/(p-1)$.

Define $F_p : [0, 1] \to \mathbb{R}$ by

$$F_p(x) = \int_0^x (1 - t^p)^{-1/p} dt. \tag{1.3.1}$$

Evidently $F_2 = \arcsin$. Since F_p is strictly increasing it has an inverse, written as $\sin_p$ by analogy with the case $p = 2$, and defined on the interval $[0, \pi_p/2]$, where

$$\pi_p = 2 \int_0^1 (1 - t^p)^{-1/p} dt. \tag{1.3.2}$$

Thus $\sin_p$ is strictly increasing on $[0, \pi_p/2]$, $\sin_p(0) = 0$ and $\sin_p(\pi_p/2) = 1$. Extend it to $[0, \pi_p]$ by defining

$$\sin_p x = \sin_p(\pi_p - x) \text{ for } x \in [\pi_p/2, \pi_p]; \qquad (1.3.3)$$

further extension to $[-\pi_p, \pi_p]$ is made by oddness; and finally $\sin_p$ is extended to the whole of $\mathbb{R}$ by $2\pi_p$-periodicity. It is clear that this extension is continuously differentiable on $\mathbb{R}$.

Now define $\cos_p : \mathbb{R} \to \mathbb{R}$ by

$$\cos_p x = \frac{d}{dx} \sin_p x, \ x \in \mathbb{R}. \qquad (1.3.4)$$

Evidently $\cos_p$ is even, $2\pi_p$-periodic and odd about $\pi_p/2$. If $x \in [0, \pi_p/2]$ and we put $y = \sin_p x$, then

$$\cos_p x = (1 - y^p)^{1/p} = (1 - (\sin_p x)^p)^{1/p}.$$

Thus $\cos_p$ is strictly decreasing on $[0, \pi_p/2]$, $\cos_p(0) = 1$ and $\cos_p(\pi_p/2) = 0$. Also

$$|\sin_p x|^p + |\cos_p x|^p = 1; \qquad (1.3.5)$$

this is immediate if $x \in [0, \pi_p/2]$, and it holds for all $x \in \mathbb{R}$ in view of symmetry and periodicity. Despite this attractive property there is not a complete analogy between these p-functions and the classical trigonometric functions. For example, while the extended $\sin_p$ function is continuously differentiable on $\mathbb{R}$, if $2 < p < \infty$, then its second derivative is not continuous at $\pi_p/2$ because use of (1.3.5) shows its value at $x \in [0, \pi_p/2)$ to be $-h(\sin_p x)$, where

$$h(y) = (1 - y^p)^{\frac{2}{p}-1} y^{p-1}.$$

To calculate π_p, make the change of variable $t = s^{1/p}$ in (1.3.2). Then

$$\pi_p/2 = p^{-1} \int_0^1 (1 - s)^{-1/p} s^{1/p-1} ds = p^{-1} B(1 - 1/p, 1/p) = p^{-1}\Gamma(1 - 1/p)\Gamma(1/p),$$

where B is the Beta function. Hence

$$\pi_p = \frac{2\pi}{p\sin(\pi/p)}. \qquad (1.3.6)$$

Plainly $\pi_2 = \pi$ and

$$p\pi_p = 2\Gamma(1/p')\Gamma(1/p) = p'\pi_{p'}. \qquad (1.3.7)$$

From (1.3.6) and (1.3.7) we see that π_p decreases as p increases, with

$$\lim_{p\to 1} \pi_p = \infty, \ \lim_{p\to\infty} \pi_p = 2, \ \lim_{p\to 1}(p - 1)\pi_p = \lim_{p\to 1}\pi_{p'} = 2. \qquad (1.3.8)$$

An analogue of the tangent function is obtained by defining

$$\tan_p x = \frac{\sin_p x}{\cos_p x} \tag{1.3.9}$$

for all $x \in \mathbb{R}$ except for the points $(k + 1/2)\pi_p$ $(k \in \mathbb{Z})$. Clearly $\tan_p$ is odd and π_p-periodic; also $\tan_p 0 = 0$. Use of (1.3.5) shows that on $(-\pi_p/2, \pi_p/2)$, $\tan_p$ has derivative $1 + |\tan_p x|^p$. With the inverse of $\tan_p$ on this interval denoted by A, it follows that

$$A'(t) = 1/(1 + |t|^p), \ t \in \mathbb{R}.$$

Next we record some basic facts about derivatives of the p-trigonometric functions. They follow immediately from the definitions and (1.3.5).

Proposition 1.3.1. *For all $x \in [0, \pi_p/2)$,*

$$\frac{d}{dx} \cos_p x = - \sin_p^{p-1} x \cos_p^{2-p} x, \qquad \frac{d}{dx} \tan_p x = 1 + \tan_p^p x,$$

$$\frac{d}{dx} \cos_p^{p-1} x = -(p-1) \sin_p^{p-1} x, \qquad \frac{d}{dx} \sin_p^{p-1} x = (p-1) \sin_p^{p-2} x \cos_p x.$$

There are connections between the generalised trigonometric functions we have been discussing and some functions from classical analysis. For example, consider the incomplete Beta function $I(\cdot; a, b)$, defined for any positive a and b by

$$I(x; a, b) = \frac{1}{B(a, b)} \int_0^x t^{a-1}(1 - t)^{b-1} dt, \ x \in [0, 1];$$

see, for example, [2], 26.5.1. The change of variable $u = t^p$ in (1.3.1) shows that

$$F_p(x) = p^{-1} \int_0^{x^p} u^{-1/p'}(1 - u)^{-1/p} du = p^{-1} B(1/p, 1/p') I(x^p; 1/p, 1/p'),$$

and so, by (1.3.6),

$$\sin_p^{-1}(x) = F_p(x) = \frac{1}{2}\pi_p I(x^p; 1/p, 1/p'), \ x \in [0, 1].$$

Moreover, since the incomplete Beta function is related to the hypergeometric function F by

$$I(x; a, b) = \frac{x^a}{aB(a, b)} F(a, 1 - b; a + 1; x)$$

(see [2], 6.6.2), we have

$$\sin_p^{-1}(x) = x F(1/p, 1/p; 1 + 1/p; x^p), \ x \in [0, 1]. \tag{1.3.10}$$

Now the well-known fact that

$$F(a, b; c; x) = \sum_{n=0}^{\infty} \frac{\Gamma(a + n)\Gamma(b + n)\Gamma(c)}{\Gamma(a)\Gamma(b)\Gamma(c + n)} \frac{x^n}{n!}$$

gives the expansion

$$\sin_p^{-1}(x) = x \sum_{n=0}^{\infty} \frac{\Gamma(n+1/p)}{(pn+1)\Gamma(1/p)} \frac{x^{np}}{n!}, \quad x \in [0,1). \tag{1.3.11}$$

From this a series expansion for $\sin_p(x)$ when $x \in [0, \pi_p/2)$ may be obtained in the form $x \sum_{n=0}^{\infty} a_n x^{pn}$, the first three terms being

$$\sin_p x = x - \frac{1}{p(p+1)} x^{p+1} - \frac{(p^2 - 2p - 1)}{2p^2(p+1)(2p+1)} x^{2p+1} + \cdots .$$

The coefficients of subsequent terms are rather complicated.

Sample indefinite integrals involving the p-trigonometric functions are given now.

Proposition 1.3.2. *For all $x \in (0, \pi_p/2)$,*

$$\int \cos_p x dx = \sin_p x, \qquad p \int \cos_p^p x dx = (p-1)x + \sin_p x \cos_p^{p-1} x,$$

$$(p-1) \int \sin_p^{p-1} x dx = -\cos_p^{p-1} x, \qquad \int \tan_p^p x dx = \tan_p x - x$$

and

$$\int \sin_p x dx = \frac{1}{2} F(1/p, 2/p; 1 + 2/p; \sin_p^p x) \sin_p^2 x.$$

Proof. Only the last integral needs discussion. With the substitution $u = \sin_p x$ we have

$$\int \sin_p x dx = \int u(1-u^p)^{-1/p} du = \int u \sum_{n=0}^{\infty} \frac{\Gamma(n+1/p)}{\Gamma(1/p)} \frac{u^{pn}}{n!} du.$$

After integration the resulting series can be written in terms of the hypergeometric function. $\qquad \square$

Natural substitutions give the following elementary results for definite integrals.

Proposition 1.3.3. *Let $k, l > 0$. Then*

$$\int_0^{\pi_p/2} \sin_p^k x dx = \frac{1}{p} B\left(\frac{k+1}{p}, \frac{1}{p'}\right), \quad \int_0^{\pi_p/2} \cos_p^k x dx = \frac{1}{p} B\left(\frac{1}{p}, 1 + \frac{k-1}{p}\right)$$

and

$$\int_0^{\pi_p/2} \sin_p^k x \cos_p^l x dx = \frac{1}{p} B\left(\frac{k+1}{p}, 1 + \frac{l-1}{p}\right).$$

Further information and additional references concerning these functions and their applications can be found in [16].

As noted in the last section, $(\sin(n\pi\cdot))_{n\in\mathbb{N}}$ is a basis in $L_q(0,1)$ for any $q \in (1, \infty)$. The functions $\sin_p(n\pi\cdot)$ have a similar property, provided that p is not too close to 1: this result was first presented in [8], and we now give an account that incorporates some ideas from [16]. Define functions $f_{n,p}$ by

$$f_{n,p}(t) = \sin_p(n\pi_p t) \ (n \in \mathbb{N}, 1 < p < \infty, t \in \mathbb{R}). \tag{1.3.12}$$

When $p = 2$ these functions are simply the usual sine functions, and we write

$$e_n(t) = f_{n,2}(t) = \sin(n\pi t). \tag{1.3.13}$$

Each $f_{n,p}$ belongs to $C^1([0,1])$ and so is continuous with bounded variation on $[0,1]$: thus it has a Fourier sine expansion:

$$f_{n,p}(t) = \sum_{k=1}^{\infty} \widehat{f_{n,p}}(k) \sin(k\pi t), \ \widehat{f_{n,p}}(k) = 2 \int_0^1 f_{n,p}(t) \sin(k\pi t)dt. \tag{1.3.14}$$

As $f_{1,p}$ is symmetric about $t = 1/2$ it follows that $\widehat{f_{1,p}}(k) = 0$ when k is even and that

$$\widehat{f_{n,p}}(k) = 2 \int_0^1 f_{1,p}(nt) \sin(k\pi t)dt = 2 \sum_{m=1}^{\infty} \widehat{f_{1,p}}(m) \int_0^1 \sin(k\pi t) \sin(mn\pi t)dt$$

$$= \begin{cases} \widehat{f_{1,p}}(m) & \text{if } mn = k \text{ for some odd } m, \\ 0 & \text{otherwise.} \end{cases} \tag{1.3.15}$$

For brevity put $\tau_m(p) = \widehat{f_{1,p}}(m)$. The Fourier coefficients of the $f_{n,p}$ may be expressed in terms of the $\tau_m(p)$: we therefore focus on the behaviour of these numbers. For even m, $\tau_m(p) = 0$. If m is odd, integration by parts and the substitution $s = \cos_p(\pi_p t)$ show that

$$\tau_m(p) = 4 \int_0^{1/2} f_{1,p}(t) \sin(m\pi t)dt = \frac{4\pi_p}{m\pi} \int_0^{1/2} \cos_p(\pi_p t) \cos(m\pi t)dt$$

$$= -\frac{4\pi_p}{m^2\pi^2} \int_0^{1/2} \sin(m\pi t) \frac{d}{dt} \cos_p(\pi_p t)dt \tag{1.3.16}$$

$$= \frac{4\pi_p}{m^2\pi^2} \int_0^1 \sin\left(\frac{m\pi}{\pi_p} \cos_p^{-1} s\right) ds,$$

which immediately gives the estimate

$$|\tau_m(p)| \le 4\pi_p/(\pi m)^2 \ (m \text{ odd}). \tag{1.3.17}$$

In fact, when $1 < p < 2$ the decay of the $\tau_{2k+1}(p)$ is faster than indicated above. To establish this put $t = \cos_p^{-1} x := \phi(x)$, so that $x = \cos_p t$. Then

$$\phi'(x) = -x^{-(2-p)}(1 - x^p)^{-1/p'},$$

and so $|\phi'(x)| \geq 1$ in $(0, 1)$. Since

$$\phi''(x) = x^{p-3}(1 - x^p)^{-(2-1/p)}(2 - p - x^p),$$

we see that $\phi'(x)$ is increasing in $(0, (2 - p)^{1/p})$ and decreasing in $((2 - p)^{1/p}, 1)$. The minimum of $|\phi'(x)|$ on $(0, 1)$ is

$$\left\{ (2 - p)^{-(2-p)}(p - 1)^{-(p-1)} \right\}^{1/p} := m_p,$$

and is attained at $(2 - p)^{1/p}$. To deal with the oscillatory integral in (1.3.16) we use an estimate of van der Corput type. Put $b = (2 - p)^{1/p}$ and set $a = (2k + 1)\pi/\pi_p$. Integration by parts gives

$$\int_0^b \sin(a \cos_p^{-1} x) dx = \int_0^b \sin(a\phi(x)a\phi'(x)) \cdot \frac{1}{a\phi'(x)} dx$$

$$= \left[-\cos(a\phi(x)) \cdot \frac{1}{a\phi'(x)} \right]_0^b$$

$$+ \frac{1}{a} \int_0^b \cos(a\phi(x)) \frac{d}{dx} \{1/\phi'(x)\} dx.$$

Using the monotonicity of $\phi'(x)$ this gives

$$\left| \int_0^b \sin(a \cos_p^{-1} x) dx \right| \leq \frac{1}{am_p} + \frac{1}{a} \int_0^b \left| \frac{d}{dx} \{1/\phi'(x)\} \right| dx$$

$$= \frac{1}{am_p} + \frac{1}{a} \left| \int_0^b \frac{d}{dx} \{1/\phi'(x)\} dx \right|$$

$$= \frac{2}{am_p}.$$

Similarly we have

$$\left| \int_b^1 \sin(a \cos_p^{-1} x) dx \right| \leq \frac{2}{am_p}.$$

Hence

$$\left| \int_b^1 \sin\left(\frac{(2k+1)\pi}{\pi_p} \cos_p^{-1} s \right) ds \right| \leq \frac{4\pi_p}{(2k+1)\pi m_p}.$$

Together with (1.3.16) this shows that if $1 < p < 2$, then

$$|\tau_{2k+1}(p)| \leq \frac{16\pi_p^2}{m_p(2k+1)^3\pi^3} \quad (k \in \mathbb{N}).$$

Next we consider the dependence of $\sin_p(n\pi_p t)$ on p.

Proposition 1.3.4. *Suppose that $1 < p < q < \infty$. Then the function f defined by*

$$f(t) = \frac{\sin_q^{-1}(t)}{\sin_p^{-1}(t)}$$

is strictly decreasing on $(0,1)$.

Proof. Let

$$g(t) = \frac{(1-t^q)^{1/q}}{(1-t^p)^{1/p}} \quad (0 < t < 1).$$

For all $t \in (0,1)$,

$$g'(t) = g(t) \left\{ \frac{-t^{q-1}}{1-t^q} + \frac{t^{p-1}}{1-t^p} \right\} = \frac{(t^p - t^q)g(t)}{t(1-t^q)(1-t^p)} > 0.$$

The function G given by

$$G(t) = \sin_p^{-1}(t) - g(t)\sin_q^{-1}(t)$$

has negative gradient:

$$G'(t) = -(\sin_q^{-1} t)g'(t) < 0 \text{ in } (0,1).$$

Hence $G(t) < 0$ in $(0,1)$; thus

$$f'(t) = \frac{G(t)}{(\sin_q^{-1} t)^2(1-t^q)^{1/q}} < 0 \text{ in } (0,1). \qquad \Box$$

As a consequence we have

Corollary 1.3.5.
(i) *If $1 < p < q < \infty$, then*

$$1 > \frac{\sin_q^{-1}(t)}{\sin_p^{-1}(t)} \geq \frac{\pi_q}{\pi_p} \text{ in } (0,1].$$

(ii) *If $1 < p \le q < \infty$, then*

$$\sin_p^{-1}(t) \ge \sin_q^{-1}(t) \ and \ \frac{1}{\pi_q} \sin_q^{-1}(t) \ge \frac{1}{\pi_p} \sin_p^{-1}(t) \ in \ [0,1].$$

(iii) *If $1 < p \le q < \infty$, then*

$$\sin_p(\pi_p t) \ge \sin_q(\pi_q t) \ in \ [0, 1/2].$$

The following analogue of the classical Jordan inequality will also be useful.

Proposition 1.3.6. *Let $1 < p < \infty$. For all $\theta \in (0, \pi_p/2]$,*

$$\frac{2}{\pi_p} \le \frac{\sin_p \theta}{\theta} < 1.$$

Proof. With a change of variable we have

$$\sin_p^{-1} x = x \int_0^1 (1 - x^p s^p)^{-1/p} ds,$$

and so

$$\theta = (\sin_p \theta) \int_0^1 (1 - (\sin_p \theta)^p s^p)^{-1/p} ds.$$

The result now follows since

$$1 \le \int_0^1 (1 - (\sin_p \theta)^p s^p)^{-1/p} ds \le \frac{\pi_p}{2}$$

for all $\theta \in (0, \pi_p/2]$. $\qquad\square$

Corollary 1.3.7. *For all $p \in (1, \infty)$ and all $t \in (0, 1/2)$, $\sin_p(\pi_p t) > 2t$.*

Proof. By Proposition 1.3.6, $\sin_p \theta > 2\theta/\pi_p$ if $0 < \theta < \pi_p/2$. Now put $\theta = \pi_p t$. $\quad\square$

Given any function f on $[0, 1]$, extend it to a function $\tilde{f}$ on $[0, \infty)$ by setting

$$\tilde{f}(t) = -\tilde{f}(2k - t) \ \text{for } t \in [k, k + 1], k \in \mathbb{N}.$$

Now define maps $M_m : L_q(0, 1) \to L_q(0, 1)$ $(1 < q < \infty)$ by

$$M_m g(t) = \tilde{g}(mt), \ m \in \mathbb{N}, \ t \in (0, 1),$$

and observe that $M_m e_n = e_{mn}$.

Lemma 1.3.8. *For all $m \in \mathbb{N}$ and all $q \in (1, \infty)$ the map $M_m : L_q(0, 1) \to L_q(0, 1)$ is isometric and linear.*

Proof. Let $g \in L_q(0,1)$. Then

$$\int_0^1 |M_m g(t)|^q \, dt = m^{-1} \int_0^m |\widetilde{g}(s)|^q \, ds = m^{-1} \sum_{k=1}^m \int_{k-1}^k |\widetilde{g}(s)|^q \, ds$$

$$= m^{-1} \sum_{k=1}^m \int_0^1 |g(s)|^q \, ds = \int_0^1 |g(s)|^q \, ds. \qquad \square$$

Next, define a map T by

$$T g(t) = \sum_{m=1}^\infty \tau_m M_m g(t). \tag{1.3.18}$$

Lemma 1.3.9. *Let $p, q \in (1, \infty)$. The map T is a bounded linear map of $L_q(0,1)$ to itself with $\|T\| \le \pi_p/2$. For all $n \in \mathbb{N}$, $T e_n = f_{n,p}$.*

Proof. From (1.3.15), (1.3.17) and Lemma 1.3.8 we see that

$$\|T\| \le \sum_{m=1}^\infty \frac{4\pi_p}{(2m-1)^2 \pi^2} = \pi_p/2.$$

Another application of (1.3.15) shows that

$$T e_n = \sum_{m=1}^\infty \tau_m e_{mn} = \sum_{m=1}^\infty \widehat{f_{1,p}}(m) e_{mn} = \sum_{k=1}^\infty \widehat{f_{n,p}}(k) e_k = f_{n,p}. \qquad \square$$

Lemma 1.3.10. *There exists $p_0 \in (1, 2)$ such that if $p > p_0$, then for all $q \in (1, \infty)$, $T : L_q(0,1) \to L_q(0,1)$ has a bounded inverse.*

Proof. Since M_1 is the identity map id, we have from (1.3.15) and Lemma 1.3.8 that

$$\|T - \tau_1(p) id\| \le \sum_{j=1}^\infty |\tau_{2j+1}(p)|,$$

and so the invertibility of T will follow from Theorem II.1.2 of [88] if we can show that

$$\sum_{j=1}^\infty |\tau_{2j+1}(p)| < |\tau_1(p)|. \tag{1.3.19}$$

From (1.3.17) we have, for all $p \in (1, \infty)$,

$$\sum_{j=1}^\infty |\tau_{2j+1}(p)| \le \frac{4\pi_p}{\pi^2} \left(\frac{\pi^2}{8} - 1 \right). \tag{1.3.20}$$

Moreover, by Corollary 1.3.7,

$$\tau_1(p) = 4 \int_0^{1/2} \sin_p(\pi_p t) \sin(\pi t) dt > 4 \int_0^{1/2} 2t \sin(\pi t) dt = 8/\pi^2,$$

from which (1.3.19) follows if $2 \le p < \infty$ since $\pi_p \le \pi$.

If $1 < p < 2$, then the monotonic dependence of $\sin_p(\pi_p t)$ on p given by Corollary 1.3.5 (iii) shows that

$$\tau_1(p) > 4 \int_0^{1/2} \sin^2(\pi t) dt = 1.$$

Define p_0 by

$$\pi_{p_0} = \frac{\pi^2}{4} / \left(\frac{\pi^2}{8} - 1 \right),$$

and note that if $p > p_0$, then

$$\frac{4\pi_p}{\pi^2} \left(\frac{\pi^2}{8} - 1 \right) < 1.$$

Once more we have (1.3.19).

We summarise these results in the following theorem.

Theorem 1.3.11. *The map T is a homeomorphism of $L_q(0,1)$ onto itself for every $q \in (1,\infty)$ if $p_0 < p < \infty$, where p_0 is defined by the equation*

$$\pi_{p_0} = \frac{2\pi^2}{\pi^2 - 8}. \tag{1.3.21}$$

$\square$

Remark 1.3.12. Numerical solution of equation (1.3.21) shows that p_0 is approximately equal to 1.05.

Theorem 1.3.13. *Let $p \in (p_0, \infty)$ and $q \in (1,\infty)$. Then the family $(f_{n,p})_{n \in \mathbb{N}}$ forms a Schauder basis of $L_q(0,1)$ and a Riesz basis of $L_2(0,1)$.*

Proof. Since the e_n form a basis of $L_q(0,1)$ and T is a linear homeomorphism of $L_q(0,1)$ onto itself with $T e_n = f_{p,n}$ $(n \in \mathbb{N})$, it follows from Proposition 1.2.12 that the $f_{n,p}$ form a Schauder basis of $L_q(0,1)$. When $q = 2$, the e_n form a Riesz basis of $L_2(0,1)$, so that the map $(a_n) \longmapsto \sum_{n=1}^{\infty} a_n e_n$ is an isomorphism of l_2 onto $L_2(0,1)$. As $\sum_{n=1}^{\infty} a_n e_n = T^{-1} \left(\sum_{n=1}^{\infty} a_n f_{p,n} \right)$ and T is a homeomorphism, it follows that $(a_n) \longmapsto \sum_{n=1}^{\infty} a_n f_{p,n}$ is an isomorphism of l_2 onto $L_2(0,1)$. $\square$

The condition $p > p_0 > 1$ in this theorem arises from the techniques used in the proof: a discussion of this is given in [16]. Whether the result remains true for all $p > 1$ appears to be unknown at the moment.

Additional interesting generalisations of the classical trigonometric functions have been made by means of the function $F_{p,q}$ defined for all $p, q \in (1, \infty)$ and all $x \in [0, 1]$ by

$$F_{p,q}(x) = \int_0^x (1 - t^q)^{-1/p} dt. \tag{1.3.22}$$

The inverse of this strictly increasing function is denoted by $\sin_{p,q}$ and is defined on the interval $[0, \pi_{p,q}/2]$, where

$$\pi_{p,q} = 2 \int_0^1 (1 - t^q)^{-1/p} dt = 2q^{-1} B(1/p', 1/q). \tag{1.3.23}$$

This inverse is extended to $[0, \pi_{p,q}]$ by symmetry about $\pi_{p,q}/2$, to $[-\pi_{p,q}, \pi_{p,q}]$ by oddness and then to $\mathbb{R}$ by $2\pi_{p,q}$-periodicity. Plainly $\sin_{p,p} = \sin_p$. Now define $\cos_{p,q}$ to be the derivative of $\sin_{p,q}$. Much as in the proof of (1.3.5) it can be shown that for all $x \in \mathbb{R}$,

$$|\sin_{p,q} x|^q + |\cos_{p,q} x|^p = 1. \tag{1.3.24}$$

Some properties of the numbers $\pi_{p,q}$ will be useful. Immediately from the definition it can be seen that

for each fixed $q \in (1, \infty)$, $p \longmapsto \pi_{p,q}$ is decreasing on $(1, \infty)$ and $\lim_{p \to \infty} \pi_{p,q} = 2$,
$$\tag{1.3.25}$$

and

for each fixed $p \in (1, \infty)$, $q \longmapsto \pi_{p,q}$ is decreasing on $(1, \infty)$ and $\lim_{q \to \infty} \pi_{p,q} = 2$.
$$\tag{1.3.26}$$

Further information is provided by the next lemma.

Lemma 1.3.14. *Let $p, q \in (1, \infty)$. Then*

(i) $q\pi_{p,q} = p'\pi_{q',p'}$;
(ii) $\pi_{p,q} \leq \pi_{q',q}$ *if $p' \leq q$;*
(iii) $q\pi_{p,q} \leq p'\pi_{p,p'}$ *if $p' > q$.*

Proof. (i) Making the change of variables $y = (1 - t^q)^{1/p'}$ in (1.3.23) we see that

$$\pi_{p,q} = \frac{2p'}{q} \int_0^1 (1 - y^{p'})^{-1/q'} dy = \frac{p'}{q} \pi_{q',p'}.$$

(ii), (iii) Define $P : (0, 1)^2 \to (0, \infty)$ by $P(x, y) = \pi_{1/x, 1/y}$. By (1.3.25) and (1.3.26) we know that $P(x, y)$ is increasing in x for any fixed $y \in (0, 1)$, and increasing in y for any fixed $x \in (0, 1)$. Hence

$$P(x, y) \leq P(1 - y, y) \text{ if } y \leq 1 - x. \tag{1.3.27}$$

If $y > 1 - x$, then (i) and monotonicity give

$$P(x, y) = \frac{y}{1-x} P(1 - y, 1 - x) \le \frac{y}{1-x} P(x, 1 - x).\qquad (1.3.28)$$

Now put $x = 1/p$ and $y = 1/q$ in (1.3.27) and (1.3.28) to obtain (ii) and (iii). $\quad\square$

Next we claim that the area A_p enclosed by the p-circle $|x|^p + |y|^p = 1$ is given by

$$A_p = \pi_{p',p} \quad (1 < p < \infty).\qquad (1.3.29)$$

For

$$A_p = 4 \int\int dx dy,$$

where the integration is over all those non-negative values of x and y such that $x^p + y^p \le 1$. Put $x = w^{1/p}, y = z^{1/p}$: then

$$A_p = 4p^{-2} \int\int (wz)^{-1/p'} dw dz,$$

where the integration is over $\{(w, z) : w \ge 0, z \ge 0, w + z \le 1\}$. The integral here is of Dirichlet form (see [86], 12.5) and so

$$A_p = \frac{4(\Gamma(1/p))^2}{p^2 \Gamma(2/p)} \int\limits_0^1 \tau^{2/p-1} d\tau,$$

which establishes the claim.

For $p \in [1, \infty]$ let

$$S_p := \left\{ (x, y) \in \mathbb{R}^2 : |x|^p + |y|^p \le 1 \right\} \text{ if } p < \infty,$$

and

$$S_\infty := \left\{ (x, y) \in \mathbb{R}^2 : \max(|x|, |y|) \le 1 \right\};$$

let $|S_p|$ be the 2-Lebesgue measure of S_p. Use of the inequalities

$$\max(|x|, |y|) \le (|x|^p + |y|^p)^{1/p} \le |x| + |y| \quad (1 < p < \infty)$$

shows that

$$S_1 \subset S_p \subset S_\infty \quad (1 < p < \infty).$$

Hence

$$|S_1| \le A_p \le |S_\infty|.$$

As $|S_1| = 2$ and $|S_\infty| = 4$ this justifies the following

Lemma 1.3.15. *For all $p \in (1, \infty)$,*

$$2 \leq \pi_{p,p'} \leq 4.$$

By the same method as in the proof of Proposition 1.3.6 the following analogue of that proposition can be established.

Proposition 1.3.16. *For all $p, q \in (1, \infty)$ and all $\theta \in (0, \pi_{p,q}/2]$,*

$$\frac{2}{\pi_{p,q}} \leq \frac{\sin_{p,q}\theta}{\theta} \leq 1.$$

Finally we turn to the basis properties of the $\sin_{p,q}$ functions and show that their behaviour in this respect is better than that known for the $\sin_p$ functions provided that p' and q are not too far apart. As some of the arguments are similar to those already given when $p = q$, we shall be brief. Let $p, q \in (1, \infty)$ and for each $n \in \mathbb{N}$ put

$$f_{n,p,q}(t) = \sin_{p,q}(n\pi_{p,q}t) \quad (t \in \mathbb{R}).$$

Each $f_{n,p,q}$ has a Fourier sine expansion on $[0, 1]$:

$$f_{n,p,q}(t) = \sum_{k=1}^{\infty} \widehat{f_{n,p,q}}(k)\sin(k\pi t),$$

where arguments similar to those for the case $p = q$ show that

$$\widehat{f_{n,p,q}}(k) = \begin{cases} \widehat{f_{1,p,q}}(m) & \text{if } k = mn \text{ and } m \text{ is odd,} \\ 0 & \text{otherwise.} \end{cases}$$

Now set

$$\tau_m(p, q) = \widehat{f_{1,p,q}}(m), e_n(t) = \sin(n\pi t).$$

As before, extend any function $f : [0, 1] \to \mathbb{R}$ to a function $\widetilde{f}$ on $[0, \infty)$ by setting $\widetilde{f}(t) = -\widetilde{f}(2k - t)$ for $t \in [k, k+1]$, $k \in \mathbb{N}$ and define $M_m : L_r(0, 1) \to L_r(0, 1)$ $(m \in \mathbb{N}, r \in (1, \infty))$ by

$$M_m g(t) = \widetilde{g}(mt).$$

Then M_m is a linear isometry and the map T given by

$$Tg(t) := \sum_{m=1}^{\infty} \tau_m(p, q) M_m g(t)$$

is a bounded linear map of $L_r(0, 1)$ to itself such that $T e_n = f_{n,p,q}$ for all $n \in \mathbb{N}$. To prove that the $f_{n,p,q}$ form a basis of $L_r(0, 1)$ it is sufficient to show that T is a homeomorphism, and this will be the case if

$$\sum_{m=1}^{\infty} |\tau_{2k+1}(p, q)| < |\tau_1(p, q)|. \tag{1.3.30}$$

As in (1.3.16) we have

$$\tau_{2k+1}(p,q) = \frac{4\pi_{p,q}}{(2k+1)^2\pi^2} \int_0^1 \sin\left(\frac{(2k+1)\pi}{\pi_{p,q}}(\cos_{p,q})^{-1}(s)\right) ds,$$

which gives

$$\sum_{m=1}^{\infty} |\tau_{2k+1}(p,q)| \leq \frac{\pi_{p,q}(\pi^2-8)}{2\pi^2}. \tag{1.3.31}$$

To estimate $|\tau_1(p,q)|$ from below, note that by virtue of Proposition 1.3.16, $\sin_{p,q}(\pi_{p,q}t) \geq 2t$ for all $t \in (0, 1/2)$. Hence

$$\tau_1(p,q) \geq 4 \int_0^{1/2} 2t\sin(\pi t)dt = 8/\pi^2. \tag{1.3.32}$$

The combination of (1.3.31) and (1.3.32) thus shows that $(\sin_{p,q}(n\pi_{p,q}t))_{n\in\mathbb{N}}$ is a basis of every $L_r(0,1)$ $(1 < r < \infty)$ if

$$\pi_{p,q} < \frac{16}{\pi^2-8}. \tag{1.3.33}$$

Finally we have

Theorem 1.3.17. *Let $p, q \in (1, \infty)$ and suppose that*

$$\frac{p'}{q} < \frac{4}{\pi^2-8}. \tag{1.3.34}$$

Then $(\sin_{p,q}(n\pi_{p,q}t))_{n\in\mathbb{N}}$ is a basis of every $L_r(0,1)$ $(1 < r < \infty)$.

Proof. We simply have to verify (1.3.33). If $p' \leq q$, then by Lemma 1.3.14 (ii) and Lemma 1.3.15, $\pi_{p,q} \leq 4$ and (1.3.33) holds. If $p' > q$, Lemma 1.3.14 (iii) and Lemma 1.3.15 give $\pi_{p,q} \leq 4p'/q$ so that again (1.3.33) holds. ☐

The number $4/(\pi^2-8)$ is approximately equal to 2.14. A particular case of the last theorem is that for all $p \in (1, \infty)$ the functions $(\sin_{p,p'}(n\pi_{p,p'}t))_{n\in\mathbb{N}}$ form a basis of $L_r(0,1)$ whenever $1 < r < \infty$. In this respect their properties are superior to those known for the $\sin_p$ functions. That the pair (p, p') of indices may be somewhat special has also been indicated by the formula (1.3.29) for the area enclosed by the p-circle.

1.4   Entropy numbers and s-numbers

1.4.1   Fundamentals

The s-numbers described below are extremely useful in estimating the eigenvalues of operators acting between Banach spaces. Let $\mathcal{B}$ be the family of all bounded linear operators acting between arbitrary Banach spaces. A map $s = (s_n) : \mathcal{B} \to [0,\infty)^{\mathbb{N}}$ which to each $T \in \mathcal{B}$ assigns a sequence $(s_n(T))$ of nonnegative real numbers is called an s-**function** if, for all Banach spaces W, X, Y and Z, it has the following properties:

(S1) $\|S\| = s_1(S) \geq s_2(S) \geq \cdots \geq 0$ for all $S \in B(X,Y)$;

(S2)   for all $S_1, S_2 \in B(X,Y)$ and all $n \in \mathbb{N}$,

$$s_n\left(S_1 + S_2\right) \leq s_n\left(S_1\right) + \|S_2\|\,;$$

(S3)   for all $S \in B(X,Y)$, $R \in B(Y,Z)$ and $U \in B(Z,W)$ and all $n \in \mathbb{N}$,

$$s_n(URS) \leq \|U\|\, s_n(R)\, \|S\|\,;$$

(S4)   for all $S \in B(X,Y)$ with rank $S < n \in \mathbb{N}$,

$$s_n(S) = 0;$$

(S5) $s_n(I_n) = 1$ for all $n \in \mathbb{N}$, where I_n is the identity map of

$$l_2^n := \{x \in l_2 : x_j = 0 \text{ if } j > n\}$$

to itself.

For all $n \in \mathbb{N}$, $s_n(S)$ is called the nth s-**number** of S. All s-numbers coincide for operators acting between Hilbert spaces; this is not so outside the Hilbert space framework. Note also that using (S3) and (S5) the converse of (S4) can be established:

$$S \in B(X,Y), \ \ s_n(S) = 0 \text{ implies that rank } S < n.$$

For if rank $S \geq n$, then $I_n = R \circ S \circ T$ for some $T \in B(l_2^n, X)$ and $R \in B(Y, l_2^n)$. Thus $1 = s_n(I_n) \leq \|R\| s_n(S) \|T\| = 0$ and we have a contradiction.

An s-function is called **additive** if for all $m, n \in \mathbb{N}$ and all $S_1, S_2 \in B(X,Y)$, where X and Y are arbitrary Banach spaces,

$$s_{m+n-1}\left(S_1 + S_2\right) \leq s_m(S_1) + s_n(S_2);$$

it is said to be **multiplicative** if for all $m, n \in \mathbb{N}$ and all $S \in B(X,Y)$ and $R \in B(Y,Z)$, where X, Y and Z are arbitrary Banach spaces,

$$s_{m+n-1}(RS) \leq s_m(R)s_n(S).$$

Given any Banach spaces X and Y, a linear map $J : X \to Y$ such that $\|Jx\| = \|x\|$ for all $x \in X$ is called a **metric injection**; a map $Q : X \to Y$ that maps the open unit ball of X onto the open unit ball of Y is called a **metric surjection**. An s-function $s = (s_n)$ is said to be **injective** if, given any metric injection $J : Y \to Y_0$, $s_n(JT) = s_n(T)$ for all $T \in B(X,Y)$ and all Banach spaces X, so that $s_n(T)$ does not depend on the size of the target space Y; the s-function is said to be **surjective** if, given any metric surjection $Q : X_0 \to X$, $s_n(TQ) = s_n(T)$ for all $T \in B(X,Y)$ and all Banach spaces Y, so that $s_n(T)$ does not depend on the size of the domain space X.

For the basic facts about these numbers which are given below we refer to the books of Pietsch ([72]–[74]). A list of commonly used s-numbers is given below for arbitrary $S \in B(X,Y)$ and $n \in \mathbb{N}$:

(i) the **approximation numbers** $a_n(S)$,

$$a_n(S) := \inf\{\|S - F\| : F \in B(X,Y),\ \mathrm{rank}\,F < n\};$$

(ii) the **Kolmogorov numbers** $d_n(S)$,

$$d_n(S) := \inf\left\{\|Q_M^Y S\| : M \text{ is a linear subspace of } Y, \dim M < n\right\},$$

where Q_M^Y is the canonical map of Y onto Y/M;

(iii) the **Gelfand numbers** $c_n(S)$,

$$c_n(S) := \inf\left\{\|SJ_M^X\| : M \text{ is a linear subspace of } X, \mathrm{codim}\,M < n\right\},$$

where J_M^X is the embedding map from M to X;

(iv) the **Bernstein numbers** $b_n(S)$,

$$b_n(S) := \sup\left\{\inf_{x \in X_n \setminus \{0\}} \frac{\|Sx\|_Y}{\|x\|_X} : X_n \text{ is a linear subspace of } X, \dim X_n \geq n\right\};$$

(v) the **Weyl numbers** $w_n(S)$,

$$w_n(S) := \sup\{a_n(SA) : \|A : l_2 \to X\| \leq 1\};$$

(vi) the **Hilbert numbers** $h_n(S)$,

$$h_n(S) := \sup\{a_n(VSU) : \|U : l_2 \to X\| \leq 1, \|V : Y \to l_2\| \leq 1\}.$$

The approximation, Kolmogorov, Gelfand and Weyl numbers form additive and multiplicative s-functions; the Bernstein numbers are not multiplicative (see [75]); the approximation numbers are the largest s-numbers and the Hilbert numbers the smallest; the Gelfand numbers are the largest injective s-numbers and the Kolmogorov numbers are the largest surjective s-numbers; the Weyl numbers are injective. Moreover, for all $S \in B(X,Y)$ and all $n \in \mathbb{N}$,

$$\left.\begin{array}{l} b_n(S) \leq \min\{c_n(S), d_n(S)\}, w_n(S) \leq c_n(S), \\[4pt] c_{2n-1}(S) \leq 2e\sqrt{n}\left(\prod_{k=1}^n w_k(S)\right)^{1/n}, \\[4pt] w_{2n-1}(S) \leq \sqrt{n}\left(\prod_{k=1}^n h_k(S)\right)^{1/n}, \\[4pt] a_n(S) \leq 2n^{1/2}\min\{c_n(S), d_n(S)\}. \end{array}\right\} \tag{1.4.1}$$

If X is a Hilbert space, then $w_n(S) = c_n(S) = a_n(S)$; if instead Y is a Hilbert space, then $d_n(S) = a_n(S)$.

Since for all bounded linear maps and any s-function the sequence $(s_n(S))$ is non-increasing and bounded below by 0, it has a limit. In particular, $\alpha(S) :=$ $\lim_{n\to\infty} a_n(S)$ exists. If $\alpha(S) = 0$, then $S : X \to Y$ is compact; but the converse is false, in general. However, suppose the target space Y has the approximation property. Then (see Theorem 1.2.25) compact linear maps from X to Y can be approximated arbitrarily closely by finite-dimensional maps, and so $S : X \to Y$ is compact if and only if $\alpha(S) = 0$. This characterisation of compactness thus holds if the target Y is an L_p space with $1 \le p < \infty$. In contrast, no matter what the target space is, an operator T is compact if and only if $c_n(T) \to 0$; the same holds with $c_n(T)$ replaced by $d_n(T)$.

In addition to the s-numbers, there are the **entropy numbers**, which play a most useful part in determining the compactness properties of maps. Given any $n \in \mathbb{N}$, the nth entropy number $e_n(S)$ of a map $S \in B(X, Y)$ is defined to be

$$\inf\{\varepsilon > 0 : S(B_X) \text{ can be covered by } 2^{n-1} \text{ balls in } Y \text{ with radius } \varepsilon\}.$$

These numbers are monotonic decreasing as n increases, and also have the additive and multiplicative properties mentioned above; they are not s-numbers as they do not have property ($S4$). In fact, if X is real, with dim $X = m < \infty$, and $I : X \to X$ is the identity map, then for all $n \in \mathbb{N}$,

$$1 \le 2^{(n-1)/m} e_n(I) \le 4. \tag{1.4.2}$$

For accounts, with many references, of these assertions and most of those given below see [21] and [74]. Connections between the entropy numbers and s-numbers are given by the inequalities

$$\max\{c_n(S), d_n(S)\} \le n e_n(S), \; x_n(S) \le 2\sqrt{n} e_n(S). \tag{1.4.3}$$

No pointwise inequality of the form $e_n(S) \le C a_n(S)$ can be expected, in general: see [21], p. 106 for an example of a diagonal map $D : l_p \to l_p$ $(1 < p < \infty)$ such that $a_n(D) = 2^{-n}$ and

$$2^{-1-\sqrt{2n}} \le e_{n+1}(D) \le 3 \cdot 2^{-\sqrt{2n}+1/2}.$$

However, the desired inequality does hold if the approximation numbers do not decrease too quickly: it is valid if there is a positive constant c such that for all $n \in \mathbb{N}$, $a_{2^{n-1}}(S) \le c a_{2^n}(S)$ (see [42], p. 15).

Clearly $\lim_{n\to\infty} e_n(S)$ exists for all $S \in B(X, Y)$. This limit is denoted by $\beta(S)$ and is called the (**ball-**) **measure of noncompactness** of S, a name that is warranted since $\beta(S) = 0$ if and only if S is compact. In general, $\beta(S) \ne \alpha(S)$, but for the embedding id of the Sobolev space $W_p^1(\Omega)$ in $L_p(\Omega)$ $(1 < p < \infty)$, under mild conditions on the open subset Ω of $\mathbb{R}^n$, we have $\beta(\text{id}) = \alpha(\text{id})$ (see [32], V.5).

Now we turn to connections between these numbers and eigenvalues of compact linear maps acting in a (complex) Banach space. Let $S \in B(X)$ be compact. Its spectrum, apart from the point 0, consists solely of eigenvalues of finite algebraic multiplicity: let $(\lambda_n(S))$ be the sequence of all nonzero eigenvalues of S, repeated according to their algebraic multiplicity and ordered by decreasing modulus; if S has only m $(< \infty)$ distinct eigenvalues and M is the sum of their algebraic multiplicities, put $\lambda_n(S) = 0$ for all $n \in \mathbb{N}$ with $n > M$. For s-numbers, there are striking results usually referred to as Weyl inequalities. In additive form these include the inequality (see [59])

$$\left(\sum_{k=1}^{n} |\lambda_k(S)|^p \right)^{1/p} \leq c_p \left(\sum_{k=1}^{n} a_k(S)^p \right)^{1/p} \quad (0 < p < \infty), \tag{1.4.4}$$

where $c_p > 1$ is a constant depending only on p. On the multiplicative side, Pietsch [73] showed that

$$\left(\prod_{k=1}^{2n-1} |\lambda_k(S)| \right)^{1/(2n-1)} \leq \sqrt{2e} \left(\prod_{k=1}^{n} w_k(S) \right)^{1/n} \quad (n \in \mathbb{N}). \tag{1.4.5}$$

Another multiplicative inequality is that recently proved by Carl [19]: for all $a > 1$ and all $n \in \mathbb{N}$,

$$\left(\prod_{k=1}^{n} |\lambda_k(S)| \right)^{1/n} \leq \sqrt{a} \left(\prod_{k=1}^{N(n)} s_k(S) \right)^{1/N(n)}, \tag{1.4.6}$$

where $N(n)$ is the integer part of $n/(1 + \log_a n)$ and the s_k are any multiplicative s-numbers.

A most interesting connection between entropy numbers and eigenvalues was discovered by Carl [17] in the form

$$|\lambda_n(S)| \leq \left(\sqrt{2} \right)^{(k-1)/n} e_k(S) \quad (k, n \in \mathbb{N}), \tag{1.4.7}$$

showing that in particular,

$$|\lambda_n(S)| \leq \sqrt{2} e_{n+1}(S) \quad (n \in \mathbb{N}). \tag{1.4.8}$$

This was subsequently extended by Carl and Triebel [22] to give

$$\left(\prod_{k=1}^{n} |\lambda_k(S)| \right)^{1/n} \leq \inf_{k \in \mathbb{N}} 2^{k/(2n)} e_k(S). \tag{1.4.9}$$

1.4.2   Gelfand numbers and widths

Let X and Y be Banach spaces and suppose that $T \in B(X,Y)$. The Gelfand numbers $c_n(T)$ of T defined in the last section are given by

$$c_n(T) := \inf\{\|TJ_M^X\| : \operatorname{codim} M < n\} \ (n \in \mathbb{N}),$$

where J_M^X is the natural embedding from the closed linear subspace M of X into X; an equivalent formulation is

$$c_n(T) = \inf_{x_1^*,\dots,x_{n-1}^* \in X^*} \sup\{\|Tx\|_Y : x \in B_X, \langle x, x_k^* \rangle = 0 \text{ for } k < n\}. \qquad (1.4.10)$$

In connection with the series representation of compact linear maps to be discussed in Chapter 2 we need the Gelfand **widths** $\tilde{c}_n(T)$ of T, defined by

$$\tilde{c}_n(T) = \inf_{L_n} \sup\{\|Tx\|_Y : \|x\|_X \leq 1, Tx \in L_n\} \ (n \in \mathbb{N}),$$

where the infimum is taken over all closed linear subspaces L_n of Y with codimension at most $n-1$, or equivalently by

$$\tilde{c}_n(T) = \inf_{y_1^*,\dots,y_{n-1}^* \in Y^*} \sup\{\|Tx\|_Y : x \in B_X, \langle Tx, y_k^* \rangle = 0 \text{ for } k < n\}. \qquad (1.4.11)$$

While the Gelfand numbers are injective but not surjective, the Gelfand widths are both injective and surjective. The nth Gelfand width coincides with the nth Gelfand number if T^* is surjective, but in general we have only $c_n(T) \leq \tilde{c}_n(T)$ for all $n \in \mathbb{N}$. Moreover, as pointed out by Pietsch ([74], p. 336), the two sets of numbers are not equivalent. He gives an example in which $T_n = I_n \circ Q_n : l_1 \to l_\infty^n$, where $I_n : l_2^n \to l_\infty^n$ is the natural embedding and $Q_n : l_1 \to l_2^n$ is any metric surjection. In [20] the following penetrating analysis of the position is given. Since the $\tilde{c}_n(T_{2n})$ are surjective and I_n^* is surjective, results of Stečkin and Pietsch (see [74], p. 334) show that

$$\begin{aligned}
\tilde{c}_n(T_{2n}) = \tilde{c}_n(I_{2n} : l_2^{2n} \to l_\infty^{2n}) &= c_n(I_{2n} : l_2^{2n} \to l_\infty^{2n}) \\
&= a_n(I_{2n} : l_2^{2n} \to l_\infty^{2n}) \\
&= \sqrt{1 - \frac{n-1}{2n}} = \sqrt{\frac{n+1}{2n}} \geq \frac{1}{\sqrt{2}}.
\end{aligned} \qquad (1.4.12)$$

However, if $T \in B(X,Y)$, then (see [74], 6.2.3.8 and 4.9.2.1-6) $c_n(T) = a_n(T)$ if Y is an l_∞ space, and $d_n(T) = a_n(T)$ if X is an l_1 space. Thus

$$c_n(T_{2n}) = a_n(T_{2n}) = d_n(T_{2n}) = d_n(I_{2n} : l_2^{2n} \to l_\infty^{2n}) \approx n^{-1/2}, \qquad (1.4.13)$$

the final equivalence coming from a result of Garnaev and Gluskin (see [74], 6.2.5.2). In view of (1.4.12) and (1.4.13) it follows that the $\tilde{c}_n$ can be much bigger than the c_n and even larger than the a_n. Thus the $\tilde{c}_n(T)$ are not s-numbers.

We now show (following [40]) that $\tilde{c}_n(T) = c_n(T)$ for all $n \in \mathbb{N}$ when X and Y are uniformly convex and uniformly smooth real Banach spaces and T has trivial kernel and range dense in Y. **For the remainder of this section it will be assumed that X, Y and T have these properties.** Given any non-empty, bounded, closed subsets A, B of X, we denote by $\delta(A, B)$ the Hausdorff distance between them:

$$\delta(A, B) := \max \left\{ \sup_{x \in A} \inf_{y \in B} \|x - y\| , \sup_{y \in B} \inf_{x \in A} \|x - y\| \right\}.$$

The function δ is a metric on the space of all such subsets. We shall also need the distance between closed linear subspaces M, N of X defined by

$$d(M, N) = \max \left\{ \sup_{x \in M \cap S_X} \inf_{y \in N} \|x - y\| , \sup_{y \in N \cap S_X} \inf_{x \in M} \|x - y\| \right\}.$$

This is equivalent to

$$\tilde{d}(M, N) := \delta \left(M \cap S_X, N \cap S_X \right);$$

in fact it is easy to see that

$$d(M, N) \leq \tilde{d}(M, N) \leq \frac{2d(M, N)}{1 + d(M, N)} \leq 2d(M, N). \tag{1.4.14}$$

We observe that

$$d(M, N) \leq \delta \left(M \cap B_X, N \cap B_X \right) \leq \tilde{d}(M, N). \tag{1.4.15}$$

For

$$\sup_{x \in M \cap S_X} \inf_{y \in N} \|x - y\| = \sup_{x \in M \cap B_X} \inf_{y \in N} \|x - y\| \leq \sup_{x \in M \cap B_X} \inf_{y \in N \cap B_X} \|x - y\|,$$

from which, and the companion inequality with M and N interchanged, the left-hand inequality in (1.4.15) follows. Similar considerations give the right-hand inequality. Note also that

$$d(M, N) = d(M^0, N^0), \tag{1.4.16}$$

where M^0, N^0 denote the polars of M, N respectively.

We begin with an immediate consequence of Theorem 1.1.33.

Lemma 1.4.1. *Let $z^* \in S_{X^*}$ and denote by Z the polar of $\{z^*\}$. Then there exists $z \in S_X$ such that $\langle z, z^* \rangle = 1$ and $z \perp^j Z$. Moreover, each $x \in X$ may be uniquely decomposed as $x = x_1 + x_2$, where $x_1 \in \mathrm{sp}\, z$, $x_2 \in Z$ and $\|x\| \geq \|x_1\| = \mathrm{dist}\,(x, Z)$.*

The next lemma shows that if two points on the unit sphere of X^* are close together, then so are those parts of their polar sets that lie in the unit ball of X.

Lemma 1.4.2. *Let $\varepsilon > 0$ and suppose that $s^*, z^* \in S_{X^*}$ are such that*

$$\|s^* - z^*\|_{X^*} < \varepsilon/4;$$

let S, Z be the polars of $\{s^*\}, \{z^*\}$ respectively. Then

$$\delta\left(S \cap B_X, Z \cap B_X\right) < \varepsilon.$$

Proof. Suppose that $\delta(S \cap B_X, Z \cap B_X) \geq \varepsilon$. Then either there exists $x \in S \cap B_X$ such that $\operatorname{dist}(x, Z \cap B_X) > \varepsilon/2$, or there exists $x \in Z \cap B_X$ such that $\operatorname{dist}(x, S \cap B_X) > \varepsilon/2$; without loss of generality suppose the second is the case. By Lemma 1.4.1, $X = \operatorname{sp}\{s\} \oplus S$ for some $s \in S_X$, and so $x = x_1 + x_2$ for some $x_1 \in \operatorname{sp}\{s\}$ and $x_2 \in S$, with $\|x\| \geq \|x_1\| = \operatorname{dist}(x, S)$. Thus x_2 is the element of S closest to x. Note that $\|x_1\| \leq 1$ and $\operatorname{dist}(x, S) \leq \operatorname{dist}(x, S \cap B_X)$. If $\|x_2\| \leq 1$, then $x_2 \in S \cap B_X$ and

$$\|x_1\| = \operatorname{dist}(x, S) = \operatorname{dist}(x, S \cap B_X).$$

On the other hand, if $\|x_2\| > 1$, then since $\|x_2\| \leq 1 + \|x_1\|$ and x_2 is the element of S closest to x, there exists $s \in S \cap B_X$ such that $\|s - x_2\| \leq \|x_1\|$. Thus $\|x - s\| \leq \|x_1\| + \|x_2 - s\| \leq 2\|x_1\|$, so that $\operatorname{dist}(x, S \cap B_X) \leq 2\|x_1\|$. It follows that in both cases,

$$\operatorname{dist}(x, S) \leq \operatorname{dist}(x, S \cap B_X) \leq 2\operatorname{dist}(x, S).$$

Use of Lemma 1.4.1 again now shows that

$$\langle x, s^* - z^* \rangle = \langle x, s^* \rangle = \langle x_1, s^* \rangle = \|x_1\| \langle s, s^* \rangle = \|x_1\| = \operatorname{dist}(x, S)$$

$$\geq \frac{1}{2} \operatorname{dist}(x, S \cap B_X) > \varepsilon/4.$$

Hence $\|s^* - z^*\|_{X^*} > \varepsilon/4$ and we have a contradiction. The lemma follows. $\qquad\square$

It is plain from the definitions that $c_n(T) = \tilde{c}_n(T)$ when $n = 1$. The next lemma shows that this is also true for $n = 2$.

Lemma 1.4.3. *The second Gelfand number of T coincides with the second Gelfand width:*

$$c_2(T) = \tilde{c}_2(T).$$

Proof. Let $\varepsilon > 0$. Given any $z^* \in X^*$, there exists $x_\varepsilon^* \in T^*(Y^*)$ such that $\|z^* - x_\varepsilon^*\|_{X^*} < \varepsilon$; let Z and X_ε be the polars of $\{z^*\}$ and $\{x_\varepsilon^*\}$ respectively. By Lemma 1.4.2,

$$\delta\left(Z \cap B_X, X_\varepsilon \cap B_X\right) < 2\varepsilon.$$

Hence

$$\sup_{x \in Z \cap B_X} \|Tx\| = \sup\left\{\|T(x + y - y)\| : x \in Z \cap B_X, x + y \in X_\varepsilon \cap B_X, \|y\| < 4\varepsilon\right\}$$

$$\leq \sup\left\{\|T(x + y)\| + \|Ty\| : x \in Z \cap B_X, x + y \in X_\varepsilon \cap B_X, \|y\| < 4\varepsilon\right\}$$

$$\leq \sup\left\{\|T(x + y)\| : x + y \in X_\varepsilon \cap B_X\right\} + 4\varepsilon \|T\|.$$

Thus $\tilde{c}_2(T) \leq c_2(T) + 4\varepsilon\,\|T\|$, so that $\tilde{c}_2(T) \leq c_2(T)$. As we already know the reverse inequality, the proof is complete. $\qquad\square$

Lemma 1.4.4. *Let $n \in \mathbb{N}\backslash\{1\}$ and suppose that $s_1^*, \ldots, s_n^*, z^* \in S_{X^*}$, with $s_1^*, \ldots, s_n^*$ linearly independent; let S_i, Z be the polars of $\{s_i^*\}, \{z^*\}$ respectively. Then there exists $a > 0$ such that if $\|s_n^* - z^*\| < \varepsilon$, then*

$$\delta\left((\cap_{i=1}^n S_i) \cap B_X, (\cap_{i=1}^{n-1} S_i) \cap Z \cap B_X\right) < a\varepsilon.$$

Proof. By (1.4.16),

$$\Lambda := d\left((\cap_{i=1}^{n-1} S_i) \cap S_n, (\cap_{i=1}^{n-1} S_i) \cap Z\right)$$
$$= d\left(\mathrm{sp}\,\{s_1^*, \ldots, s_n^*\}, \mathrm{sp}\,\{s_1^*, \ldots, s_{n-1}^*, z^*\}\right).$$

Let

$$A := \left\{(\alpha_1, \ldots, \alpha_n) : \text{each } \alpha_i \in \mathbb{R}, \sum_{i=1}^n \alpha_i s_i^* \in B_{X^*}\right\}.$$

Since the s_i^* are linearly independent, they span an n-dimensional subspace S of X^*, and as all norms on a finite-dimensional space are equivalent, $\max_{1 \leq i \leq n} |\alpha_i|$ is a norm on S equivalent to that induced on it by the norm on X^*: hence

$$b := \max_{(\alpha_1, \ldots, \alpha_n) \in A} |\alpha_n| < \infty.$$

Now let

$$M = {}^0\mathrm{sp}\,\{s_1^*, \ldots, s_n^*\}, \quad N = {}^0\mathrm{sp}\,\{s_1^*, \ldots, s_{n-1}^*, z^*\}.$$

Then

$$\Lambda = d\left(M^0, N^0\right) = \max\left(\Lambda_1, \Lambda_2\right),$$

where

$$\Lambda_1 = \sup_{x^* \in M^0 \cap S_X^*} \inf_{y^* \in N^0} \|x^* - y^*\|$$

and Λ_2 is defined similarly, with M and N interchanged. Hence

$$\Lambda_1 \leq \sup\left\{\left\|\sum_{i=1}^n \alpha_i s_i^* - \left(\sum_{i=1}^{n-1} \alpha_i s_i^* + \alpha_n z^*\right)\right\| : \sum_{i=1}^n \alpha_i s_i^* \in M^0 \cap S_{X^*}\right\}$$
$$= \sup\left\{\|\alpha_n(s_n^* - z^*)\| : \sum_{i=1}^n \alpha_i s_i^* \in M^0 \cap S_{X^*}\right\} \leq b\varepsilon.$$

In the same way it may be shown that $\Lambda_2 \leq b\varepsilon$. Thus

$$\delta\left((\cap_{i=1}^{n-1} S_i) \cap S_n \cap B_X, (\cap_{i=1}^{n-1} S_i) \cap Z \cap B_X\right) \leq 2b\varepsilon. \qquad\square$$

Corollary 1.4.5. *Let $n \in \mathbb{N}\backslash\{1\}$ and for each $i \in \{1,\dots,n\}$ suppose that $s_i^*, z_i^* \in S_{X^*}$ and let S_i, Z_i be the polars of $\{s_i^*\}, \{z_i^*\}$ respectively; assume that $\{s_1^*, \dots, s_n^*, z_1^*, \dots, z_n^*\}$ is linearly independent. There exists $c > 0$, depending on n, such that if $\|s_i^* - z_i^*\| < \varepsilon$ for all $i \in \{1, 2, \dots, n\}$, then*

$$\delta\left((\cap_{i=1}^n S_i) \cap B_X, (\cap_{i=1}^n Z_i) \cap B_X\right) \leq c\varepsilon.$$

Proof. Using the triangle inequality for the Hausdorff metric δ together with Lemma 1.4.4, we find that $\delta\left((\cap_{i=1}^n S_i) \cap B_X, (\cap_{i=1}^n Z_i) \cap B_X\right)$ is bounded above by

$$\delta\left((\cap_{i=1}^n S_i) \cap B_X, \left(\cap_{i=1}^{n-1} S_i\right) \cap Z_n \cap B_X\right)$$
$$+ \sum_{k=1}^n \delta\left((\cap_{i=1}^{n-k} S_i) \cap (\cap_{i=n-k+1}^n Z_i) \cap B_X, \left(\cap_{i=1}^{n-k-1} S_i\right) \cap \left(\cap_{i=n-k}^n Z_i\right) \cap B_X\right)$$
$$+ \delta\left(S_1 \cap (\cap_{i=2}^n Z_i \cap B_X, (\cap_{i=1}^n Z_i) \cap B_X\right)$$
$$\leq (n+2)a\varepsilon. \qquad \square$$

After this preparation we are able to establish the main result of this section.

Theorem 1.4.6. *For all $n \in \mathbb{N}$,*

$$c_n(T) = \tilde{c}_n(T).$$

Proof. We have simply to deal with the case $n > 2$. Let $\varepsilon > 0$. With the expression (1.4.10) for $c_n(T)$ in mind, let $x_1^*, \dots, x_{n-1}^* \in X^*$; we may suppose that these elements are linearly independent. Since $T^*(Y^*)$ is dense in X^*, there is a set $\{y_i^* : i = 1, \dots, n-1\} \subset Y^*$ such that, with $z_i^* := T^* y_i^*$ for each i, the set $\{x_1^*, \dots, x_{n-1}^*, z_1^*, \dots, z_{n-1}^*\} \subset X^*$ is linearly independent and $\|x_i^* - z_i^*\|_{X^*} < \varepsilon$ ($i = 1, \dots, n-1$). Let X_i, Z_i be the polars of $\{x_i^*\}, \{z_i^*\}$ respectively. Then from (1.4.11) we have

$$\tilde{c}_n(T) \leq \sup\left\{\|Tx\|_Y : x \in B_X \cap \left(\cap_{i=1}^{n-1} Z_i\right)\right\}.$$

Put

$$M^{n-1} = \left(\cap_{i=1}^{n-1} X_i\right) \cap B_X, \quad N^{n-1} = (\cap_{i=1}^n Z_i) \cap B_X.$$

By Corollary 1.4.5, $\delta\left(M^{n-1}, N^{n-1}\right) \leq c\varepsilon$. It follows that

$$\sup_{x \in N^{n-1}} \|Tx\| \leq \sup_{x \in M^{n-1}} \|Tx\| + c\varepsilon\|T\|.$$

Thus $\tilde{c}_n(T) \leq c_n(T)$ and the theorem follows. $\qquad \square$

Notes

1.1 The material in this section is quite standard: for background material and further results see [46], [62], [74] and the references contained in these books. Additional information relating to the remarkable theorem of James, mentioned in Remark 1.1.15, which characterises reflexive spaces is given in [53], [54], [55] and [70].

1.2 The literature on bases is enormous. An interesting overview of it is given in [74]; see also [66] and [46]. Theorem 1.2.15, which shows that a weak basis is a basis, is stated by Banach in the Appendix to his famous book [6]. Karlin [57] seems to be the first to sketch a proof; our treatment is based on the account given in [68]. For Theorem 1.2.19 see [50] and [56]. We refer to [74] for a discussion of the approximation property and its variants.

1.3 Much early work on p-trigonometric functions (with somewhat different definitions from that adopted here) was carried out by Lindqvist and Peetre: see, for example, [63] and [71]. Further details of the p-trigonometric functions are given in [16]. It is still unknown whether or not the $\sin_p$ functions form a basis of every $L_q(0,1)$ space if p is arbitrarily close to 1. The work in this direction involving the $\sin_{p,q}$ functions that is described in this section comes from [37]. These functions are connected with the Dirichlet problem for the p,q-Laplacian:

$$- \left(|u'|^{p-2} u' \right)' = \lambda |u|^{q-2} u \text{ on } (0, \pi_{p,q}), \ u(0) = u(\pi_{p,q}) = 0;$$

see 2.3 below for the case $p = q$. Addition formulae for the $\sin_p$ functions are essentially unknown if $p \neq 2$: no sensible expression is available that gives $\sin_p(x + y)$ in terms of $\sin_p x$ and $\sin_p y$. The same holds for the $\sin_{p,q}$ functions, except for the very special case with $q = p' = 4$, when the addition formulae for elliptic functions are used in [37] to show that for all $x \in [0, \pi_{4/3,4}/4)$,

$$\sin_{4/3,4}(2x) = \frac{2 \sin_{4/3,4} x \left(\cos_{4/3,4} x \right)^{1/3}}{\left(1 + 4 \left(\sin_{4/3,4} x \right)^4 \left(\cos_{4/3,4} x \right)^{4/3} \right)^{1/2}}.$$

In [64] functions $S_{1/p'}$ and $C_{1/p'}$ are introduced that are related to the generalised trigonometric functions given in the text by the formulae

$$S_{1/p'} = \sin_{p',p}, \quad C_{1/p'} = \cos_{p',p}^{1/(p-1)}.$$

In terms of these the above addition formula takes the rather more attractive form

$$S_{3/4}(2x) = \frac{2 S_{3/4}(x) C_{3/4}(2x)}{\left\{ 1 + 4 \left(S_{3/4}(x) C_{3/4}(x) \right)^4 \right\}^{1/2}}.$$

1.4 Comprehensive accounts of the theory of entropy and s-numbers are given in the books of Pietsch ([72], [73], [74]) and Carl and Stephani ([21]). In particular, [74] gives an interesting historical account of the development of these concepts and their interaction with other important parts of Banach space theory. For n-widths we refer to [76]: as Pietsch remarks in [74], for some time widths were favoured by the Russian school because they were more interested in measuring the degree of compactness of subsets than in an operator-theoretic approach.

Chapter 2

Representation of Compact Linear Operators

Compact linear operators have a key role in functional analysis and operator theory, with a particularly important place in the study of boundary-value problems for elliptic differential equations. They have properties which are reminiscent of linear operators acting in finite-dimensional spaces, and Theorem 1.2.25 shows a Banach space Y has the approximation property (AP) if and only if given any Banach space X and any compact map $T \in B(X, Y)$, T can be approximated arbitrarily closely in norm by a finite rank operator. Our initial objective is to establish a representation for any $x \in X$ in terms of a semi-orthogonal sequence $(x_n)_{n \in \mathbb{N}}$ determined by T, and this then yields a representation for Tx. The question of whether or not $(x_n)_{n \in \mathbb{N}}$ is a basis of X then arises, and since a Banach space with a basis must have the AP property, the importance of the AP property is exposed. Much of the second half of the chapter will be taken up by considerations of this important point.

2.1 Compact operators in Hilbert spaces

Of central importance in the theory of linear operators in Hilbert spaces is the result (with roots in the work of Hilbert in 1906) that a compact self-adjoint operator S acting on a Hilbert space H has the representation

$$Sx = \sum_{n \in \mathbb{N}} \lambda_n (x, e_n)_H e_n \tag{2.1.1}$$

for every $x \in H$, where $(\cdot, \cdot)_H$ is the inner product in H. In (2.1.1) the λ_n are eigenvalues of S and the e_n are the corresponding eigenvectors, thus $Se_n = \lambda_n e_n$. The set of eigenvalues is finite if T is of finite rank, but is otherwise countably infinite, in which case $\lim_{n \to \infty} \lambda_n = 0$. Each eigenvalue is repeated according to its multiplicity and arranged so that $|\lambda_n| \geq |\lambda_{n+1}|$ for all n. Furthermore, the sequence (e_n) is an orthonormal basis of $\ker(S)^{\perp}$, the orthogonal complement of the kernel of S in H.

Later, Erhard Schmidt showed that if H_1 and H_2 are Hilbert spaces and $T : H_1 \to H_2$ is a compact linear operator, then

$$Tx = \sum_{n \in \mathbb{N}} \lambda_n (x, e_n)_{H_1} f_n \qquad (2.1.2)$$

where the λ_n are now the eigenvalues of the positive compact self-adjoint operator $S = (T^*T)^{1/2}$ acting on H_1, $Se_n = \lambda_n e_n$ and $f_n = \lambda_n^{-1} Te_n, (\lambda_n \neq 0)$. The operator $S = (T^*T)^{1/2}$ is called the **absolute value** of T and denoted by $|T|$; its eigenvalues λ_n are called the **singular values** of T.

Our main objective in this chapter is to establish an analogue of (2.1.2) when T is a compact linear operator which maps a Banach space X into a Banach space Y. The essential features of Schmidt's result are preserved and the iterative procedure used in the proof is motivated by the standard proof of (2.1.1). In view of this, a reminder of the proof of the Schmidt representation should be helpful.

The following lemma provides the first step.

Lemma 2.1.1. *Let H be a Hilbert space with inner product $(\cdot, \cdot)$ and norm $\|\cdot\|$, and let S be a bounded self-adjoint operator on H. Then the norm $\|S\|$ of S satisfies*

$$\|S\| = \sup\{|(Sx, x)| : x \in H, \ \|x\| = 1\}. \qquad (2.1.3)$$

If S is compact and self-adjoint it has an eigenvalue λ_1, where $|\lambda_1| = \|S\|$.

Proof. Denote the right-hand side of (2.1.3) by $N(S)$. Clearly $N(S) \leq \|S\|$. Moreover, if $\lambda > 0$, then

$$4\|Sx\|^2 = (S[\lambda x + \lambda^{-1}Sx], \lambda x + \lambda^{-1}Sx) - (S[\lambda x - \lambda^{-1}Sx], \lambda x - \lambda^{-1}Sx)$$

for all $x \in H$. Therefore

$$4\|Sx\|^2 \leq N(S)\left(\|\lambda x + \lambda^{-1}Sx\|^2 + \|\lambda x - \lambda^{-1}Sx\|^2\right)$$
$$= 2N(S)\left(\lambda^2\|x\|^2 + \lambda^{-2}\|Sx\|^2\right).$$

If $Sx \neq 0$, the right-hand side attains its minimum as a function of λ when $\lambda^2 = \|Sx\|/\|x\|$, and on substituting this value we obtain

$$\|Sx\|^2 \leq N(S)\|Sx\|\|x\|$$

and hence $\|Sx\| \leq N(S)\|x\|$. This obviously holds also when $Sx = 0$. Consequently $\|S\| \leq N(S)$ and (2.1.3) is proved. It follows that there exists a sequence (x_n) in H such that $\|x_n\| = 1$ for all $n \in \mathbb{N}$, and $\lim_{n \to \infty}(Sx_n, x_n) = \lambda_1$, where $|\lambda_1| = \|S\|$. Hence

$$\|Sx_n - \lambda_1 x_n\|^2 = \|Sx_n\|^2 - 2\lambda_1(Sx_n, x_n) + |\lambda_1|^2\|x_n\|^2$$
$$\leq |\lambda_1|^2 - 2\lambda_1(Sx_n, x_n) + |\lambda_1|^2 \to 0 \qquad (2.1.4)$$

as $n \to \infty$. If S is also compact, (Sx_n) must contain a convergent subsequence $(Sx_{n(k)})$, say. Put $e_1 = \lambda_1^{-1} \lim_{k \to \infty} Sx_{n(k)}$. By (2.1.4), $x_{n(k)} \to e_1$ and so since S is continuous, $Sx_{n(k)} \to Se_1$. Thus $Se_1 = \lambda_1 e_1, \|e_1\| = 1$ and the proof is complete. $\qquad \square$

Proof of (2.1.1). Lemma 2.1.1 generates the following iterative procedure to establish the existence of eigenvalues of S. Let $H_2 = \{\mathrm{sp}\{e_1\}\}^\perp$, the orthogonal complement in H of the linear span of e_1, where $Se_1 = \lambda_1 e_1$, $|\lambda_1| = \|S\|$, and denote by S_2 the restriction of S to H_2. Then S_2 is a compact self-adjoint operator on the Hilbert space H_2 and so, by Lemma 2.1.1, if $S_2 \neq 0$, it has an eigenvalue λ_2 such that $|\lambda_2| = \|S_2\|$, and an eigenvector $e_2 \in H_2$ with $\|e_2\| = 1$. Hence

$$Se_2 = \lambda_2 e_2, \quad e_2 \perp e_1, \quad |\lambda_2| \leq \|S\| = |\lambda_1|.$$

This process can be continued indefinitely, unless at some stage $H_{m+1} := \{e_1, e_2, \ldots, e_m\}^\perp$ and $S_{m+1} : S \upharpoonright H_{m+1}$ is the zero operator. If the process does not terminate we obtain an infinite sequence (λ_n) of eigenvalues of S with $|\lambda_{n+1}| \leq |\lambda_n|$, and an orthonormal sequence of associated eigenvectors $(e_n) : Se_n = \lambda_n e_n$. In this latter case we must have $\lambda_n \to 0$ as $n \to \infty$ since otherwise $(\lambda_n^{-1} e_n)$ would be bounded and hence the compactness of S would imply that as $e_n = S(\lambda_n^{-1} e_n)$, the sequence (e_n) contains a convergent subsequence, contrary to the fact that it is orthonormal. The same argument implies that the multiplicity of each eigenvalue λ_n is finite.

Let m be the number of eigenvectors in the sequence (e_n) if it is finite, so that $S_{m+1} = 0$, and let m be an arbitrary positive number otherwise. Let $x \in H$ and put

$$y_m = x - \sum_{n=1}^{m} (x, e_n) e_n.$$

Since $(y_m, e_n) = 0, \ n = 1, 2, \ldots, m$, it follows that $y_m \in H_{m+1}$ and so

$$\|Sy_m\| = \|S_{m+1} y_m\| \leq \|S_{m+1}\| \|y_m\| = |\lambda_{m+1}| \|y_m\| \tag{2.1.5}$$

and

$$Sy_m = Sx - \sum_{n=1}^{m} \lambda_n (x, e_n) e_n. \tag{2.1.6}$$

The sequence (y_m) is bounded since

$$\|y_m\| \leq \|x\| + \left\| \sum_{n=1}^{m} (x, e_n) e_n \right\|$$

$$= \|x\| + \left(\sum_{n=1}^{m} |(x, e_n)|^2 \right)^{1/2}$$

$$\leq 2\|x\|$$

by Bessel's inequality. If $S_{m+1} = 0$ then $Sy_m = 0$ and so $Sx = \sum_{n=1}^{m}(x, e_n)e_n$. Otherwise, since $\lambda_{m+1} \to 0$ as $m \to \infty$, we see from (2.1.5) that $Sy_m \to 0$ and so from (2.1.6),

$$Sx = \sum_{n=1}^{\infty} \lambda_n(x, e_n)e_n.$$

It remains to prove that (e_n) is an orthonormal basis of $\ker(S)^{\perp}$. To this end, let $x \in H$ and define

$$M = \begin{cases} \infty & \text{if } \operatorname{rank}(S) = \infty, \\ \min\{m : S_{m+1} = 0\} & \text{otherwise.} \end{cases}$$

Then $\sum_{n=1}^{M}(x, e_n)e_n$ converges in H, and since S is continuous,

$$S\left(x - \sum_{n=1}^{M}(x, e_n)e_n\right) = Sx - \sum_{n=1}^{M}\lambda_n(x, e_n)e_n = 0,$$

on account of (2.1.1). Since $(Sx, e_n) = (x, Se_n) = \lambda_n(x, e_n)$ it follows that each e_n lies in $\ker(S)^{\perp}$ and thus if $x \in \ker(S)^{\perp}$,

$$x - \sum_{n=1}^{M}(x, e_n)e_n \in \ker(S) \cap \ker(S)^{\perp} = \{0\}.$$

This proves the assertion.   □

Proof of (2.1.2). Since $T : H_1 \to H_2$ is compact, $S^2 = (T^*T)$, and hence $S = (T^*T)^{1/2} =: |T|$ are compact self-adjoint operators on H_1. It therefore follows from above that

$$|T|x = Sx = \sum_{n=1}^{\infty} \lambda_n(x, e_n)_{H_1}e_n, \quad x \in H_1,$$

where $Se_n = \lambda_n e_n$ and (e_n) is an orthonormal set in H_1. This yields

$$T^*Tx = S^2x = \sum_{n=1}^{\infty} \lambda_n^2(x, e_n)_{H_1}e_n, \quad x \in H_1, \qquad (2.1.7)$$

the series being finite if T (and hence S) is of finite rank. On setting $f_n = \lambda_n^{-1}Te_n$ and noting that $S^2e_n = \lambda_n^2 e_n$, we obtain

$$T^*Tx = \sum_{n=1}^{\infty}(x, T^*Te_n)_{H_1}e_n = \sum_{n=1}^{\infty}(Tx, Te_n)_{H_2}e_n$$

$$= \sum_{n=1}^{\infty}\lambda_n(Tx, f_n)_{H_2}e_n \quad x \in H_1. \qquad (2.1.8)$$

We also have that (f_n) is an orthonormal sequence in H_2 since

$$\lambda_n \lambda_m (f_n, f_m)_{H_2} = (Te_n, Te_m)_{H_2} = (e_n, T^* Te_m)_{H_1} = \lambda_m^2 (e_n, e_m)_{H_1}.$$

Hence

$$
\left\| \sum_{n=1}^{\infty} \lambda_n (Tx, f_n)_{H_2} e_n \right\|^2 = \sum_{n=1}^{\infty} \lambda_n^2 |(Tx, f_n)_{H_2}|^2
$$
$$
\leq \lambda_1^2 \sum_{n=1}^{\infty} |(Tx, f_n)_{H_2}|^2 \leq \lambda_1^2 \|Tx\|^2
$$

(2.1.9)

by Bessel's inequality. As both sides of (2.1.8) depend continuously on Tx, it follows that for all $w \in \mathcal{R}(T)$,

$$T^* w = \sum_{n=1}^{\infty} \lambda_n (w, f_n)_{H_2} e_n.$$

Since H_2 is the orthogonal sum of $\overline{\mathcal{R}(T)}$ and its orthogonal complement, we have that any $y \in H_2$ can be written as $y = w + z$, where $w \in \overline{\mathcal{R}(T)}$ and $z \in \mathcal{R}(T)^{\perp}$. Hence $T^* z = 0$ and since each $f_n \in \mathcal{R}(T)$, we have for all $y \in H_2$,

$$T^* y = \sum_{n=1}^{\infty} \lambda_n (y, f_n)_{H_2} e_n.$$

(2.1.10)

Consequently, for all $x \in H_1$ and $y \in H_2$,

$$
(Tx, y)_{H_2} = (x, T^* y)_{H_1}
$$
$$
= \sum_{n=1}^{\infty} \lambda_n (y, f_n)_{H_2} (x, e_n)_{H_1} = \sum_{n=1}^{\infty} \lambda_n ((x, e_n)_{H_1} f_n, y)_{H_2},
$$

whence (2.1.2). $\qquad \square$

2.2 Compact operators in Banach spaces

2.2.1 Preliminaries

In this general setting $T : X \rightarrow Y$, is a compact linear operator and X, Y are Banach spaces which we assume throughout to be reflexive and (for ease of presentation) real, without further mention. We shall be assuming other conditions on X and Y which are sufficient to ensure that T has a representation which is analogous to that in (2.1.2), and which reduces to Schmidt's representation when X and Y are Hilbert spaces. Moreover, the representation will be in terms of a sequence (x_n) which, under appropriate conditions, is a Schauder basis of X.

In the proofs of Section 2.1, the notion of orthogonality given by the Hilbert space structure plays a crucial role, and so one is faced with an immediate and obvious problem in the general problem involving Banach spaces.

The following proposition (see [11], Chapter IV, section 5, Propositions 9 and 11) will be useful to us.

Proposition 2.2.1. *Let N be a closed linear subspace of the reflexive Banach space X and let ϕ be the canonical map of X onto the quotient space X/N. Then*

1) *the adjoint ϕ^* of ϕ is an isometric isomorphism of $(X/N)^*$ onto N^0;*
2) *$(N^0)^*$ is isometrically isomorphic to X/N, and X/N is reflexive.*

Proof. Put $Y = X/N$. For every $y^* \in Y^*$,

$$\|y^*\| = \sup_{y \in Y, \|y\| < 1} |\langle y, y^* \rangle_Y| = \sup_{x \in X, \|\phi(x)\| < 1} |\langle \phi(x), y^* \rangle_Y|$$
$$= \sup_{x \in X, \|\phi(x)\| < 1} |\langle x, \phi^*(y^*) \rangle_X|.$$

The open ball $\{y \in Y : \|y\| < 1\}$ is the canonical image of the open ball $\{x \in X : \|x\| < 1\}$: thus

$$\|y^*\| = \sup_{x \in X, \|x\| < 1} |\langle x, \phi^*(y^*) \rangle_X| = \|\phi^*(y^*)\|.$$

Hence ϕ^* is an isometry. Moreover, by Theorem 4.6-F of [83],

$$^0\overline{\phi^*(Y^*)} = \ker(\phi) = N,$$

so that $\overline{\phi^*(Y^*)} = N^0$, from which it follows (because of the isometric nature of ϕ^*) that ϕ^* maps Y^* onto N^0.

As for 2), note that X^* is reflexive and thus so is its closed linear subspace N^0, which is isometrically isomorphic to $(X/N)^*$. Hence $(X/N)^*$ is reflexive, and therefore so is X/N. The rest is clear. $\qquad\square$

Proposition 2.2.2. *Let N be a closed linear subspace of X. If X is strictly convex, so are N and X/N; if X^* is strictly convex, so are $(X/N)^*$ and N^0.*

Proof. Suppose X is strictly convex. Then clearly so is N. Since X is reflexive, the norm on X^*, and thus on N^0, is Gâteaux differentiable at all non-zero points by Proposition 1.1.20. But by Proposition 2.2.1, $(X/N)^*$ is isometrically isomorphic to N^0 and so X/N is strictly convex.

If X^* is strictly convex, then so is its closed linear subspace N^0 and thus also $(X/N)^*$. The asserted result therefore follows. $\qquad\square$

Before proceeding further, we introduce some notation and collect some necessary facts. For X^* strictly convex, set

$$\tilde{J}_X x := \operatorname{grad} \|x\|_X, \quad x \in X \setminus \{0\}.$$

We shall need the following properties:

$$\tilde{J}_X x \in X^*, \quad \|\tilde{J}_X x\|_{X^*} = 1; \tag{2.2.1}$$

given any $x \in X \setminus \{0\}$, $\tilde{J}_X x$ is the unique element of X^* which satisfies

$$\langle x, \tilde{J}_X x \rangle_X = \|x\|_X. \tag{2.2.2}$$

For (2.2.1) and (2.2.2) see Lemma 1.1.17. The fact that given any $x \in X \setminus \{0\}$, $\tilde{J}_X x$ is the unique element of X^* which satisfies (2.2.2) comes from the analogue of Proposition 1.1.3 for X^* strictly convex and Proposition 1.1.20. In the terminology of Remark 1.1.18, $\tilde{J}_X x$ is a supporting functional of B_X at x. For these properties we refer to Remark 1.1.18, Lemma 1.1.19 and Proposition 1.1.20. Note that $\tilde{J}_X$ is positively homogeneous, for if $\lambda > 0$,

$$\langle \lambda x, \tilde{J}_X(\lambda x) \rangle_X = \lambda \|x\|_X$$

and hence $\tilde{J}_X(\lambda x) = \tilde{J}_X x$. In general we have

$$\tilde{J}_X(\lambda x) = (\operatorname{sgn} \lambda) \tilde{J}_X x. \tag{2.2.3}$$

We recall that if $\mu : [0, \infty) \to [0, \infty)$ is a continuous, strictly increasing function which satisfies $\mu(0) = 0$, $\lim_{t \to \infty} \mu(t) = \infty$, then

$$J_X x := \mu(\|x\|_X) \tilde{J}_X x, \quad (x \in X \setminus \{0\}), \quad J_X(0) = 0, \tag{2.2.4}$$

is a duality map on X with gauge function μ; it satisfies

$$\langle x, J_X x \rangle_X = \|J_X x\|_{X^*} \|x\|_X, \quad \|J_X x\|_{X^*} = \mu(\|x\|_X). \tag{2.2.5}$$

Operators $\tilde{J}_Y, J_Y$ are defined similarly.

The first step in the procedure that will ultimately lead to the desired representation for T is the following proposition.

Proposition 2.2.3. *Let $T : X \to Y$ be compact and linear. Then there exists $x_1 \in X$, with $\|x_1\|_X = 1$, such that $\|T\| = \|T x_1\|_Y$.*

If X^*, Y^* are strictly convex and $T \neq 0$, then $x = x_1$ satisfies

$$T^* \tilde{J}_Y T x = \nu \tilde{J}_X x, \tag{2.2.6}$$

with $\nu = \|T\|$; in terms of duality maps, this equation has the form

$$T^* J_Y T x = \nu_1 J_X x, \quad \nu_1 = \|T\| \mu_Y(\|T\|). \tag{2.2.7}$$

Moreover, if $x \in X \setminus \{0\}$ satisfies (2.2.6) for some ν, then $0 \leq \nu \leq \|T\|$ and $\|T x\|_Y = \nu \|x\|_X$.

Proof. The proof of the first part is similar to that of Lemma 2.1.1. We may assume that $T \neq 0$ as otherwise the result is obvious. Let $\{w_k\}$ be a sequence of elements of X with $\|w_k\|_X = 1$ for all k and $\lim_{k\to\infty} \|Tw_k\|_Y = \|T\|$. Since $\{w_k\}$ is bounded and X is reflexive, there is a weakly convergent subsequence of $\{w_k\}$, still denoted by $\{w_k\}$ for convenience, with weak limit $w \in X$. As T is compact, $Tw_k \to Tw$ in Y. Thus $\|w\|_X \leq \liminf_{k\to\infty} \|w_k\|_X = 1$ and $\|Tw\|_Y = \|T\|$. Note that $w \neq 0$, for otherwise $T = 0$. Hence

$$\|T\| = \sup_{0 < \|x\|_X \leq 1} \frac{\|Tx\|_Y}{\|x\|_X} \geq \frac{\|Tw\|_Y}{\|w\|_X} \geq \|T\|.$$

Now set $x_1 = w/\|w\|_X$.

Since

$$\|T\| = \|Tx_1\|_Y = \max_{x \in X \setminus \{0\}} \frac{\|Tx\|_Y}{\|x\|_X}$$

it follows that for all $x \in X$,

$$\frac{d}{dt}\left(\frac{\|Tx_1 + tTx\|_Y}{\|x_1 + tx\|_X}\right)\bigg|_{t=0} = 0,$$

so that in terms of the duality pairings and the Gâteaux derivatives of the norms of $\|x\|_X$ and $\|Tx\|_Y$,

$$\langle Tx, \tilde{J}_Y Tx_1 \rangle_Y = \|Tx_1\|_Y \langle x, \tilde{J}_X x_1 \rangle_X,$$

and hence

$$T^* \tilde{J}_Y T x_1 = \lambda \tilde{J}_X x_1$$

with $\lambda = \|T\|$. The identity (2.2.7) follows from (2.2.4).

For the converse, let $x \in X \setminus \{0\}$ satisfy (2.2.6) for some ν. Then

$$\|Tx\|_Y = \langle Tx, \tilde{J}_Y Tx \rangle_Y = \langle x, T^* \tilde{J}_Y Tx \rangle_X = \nu \langle x, \tilde{J}_X x \rangle_X = \nu \|x\|_X$$

and $0 \leq \nu \leq \|T\|$.   □

Let X and Y be reflexive with strictly convex duals; we shall assume these conditions hereafter unless the contrary is specifically mentioned. We shall now apply Proposition 2.2.3 repeatedly in a way reminiscent of the proof of (2.1.1). Set $X = X_1, \lambda_1 = \|T\|$ and define

$$M_1 := \mathrm{sp}\{J_X x_1\}, \quad X_2 :=^0 M_1; \quad N_1 := \mathrm{sp}\{J_Y Tx_1\}, \quad Y_2 :=^0 N_1.$$

As X_2 and Y_2 are closed subspaces of reflexive spaces they are reflexive. They also have strictly convex duals, since, for example, $X_2^* = (^0M_1)^*$ is isometrically isomorphic to X_1^*/M_1 (by Proposition 2.2.1), which is strictly convex by Proposition 2.2.2. Moreover, since by Proposition 2.2.3,

$$\langle Tx, J_Y Tx_1 \rangle_Y = \nu_1 \langle x, J_X x_1 \rangle_X \quad \text{for all} \quad x \in X,$$

it follows that T maps X_2 into Y_2. The restriction T_2 of T to X_2 is thus a compact linear map from X_2 to Y_2. The conditions of Proposition 2.2.3 therefore apply for $T_2 : X_2 \to Y_2$ and, if $T_2 \neq 0$, we infer that there exists $x_2 \in X_2 \setminus \{0\}$ such that, with obvious notation,

$$\langle T_2 x, J_{Y_2} T_2 x_2 \rangle_{Y_2} = \nu_2 \langle x, J_{X_2} x_2 \rangle_{X_2} \quad \text{for all} \quad x \in X_2,$$

where $\nu_2 = \lambda_2 \mu_Y(\lambda_2)$, $\lambda_2 = \|T x_2\|_Y = \|T_2\|$. Evidently $\lambda_2 \leq \lambda_1$ and so $\nu_2 \leq \nu_1$. On identifying Y_2^* with Y^*/Y_2^0, it follows that $J_{Y_2} y - J_Y y \in Y_2^0$ for any $y \in Y_2$ and hence, if $x \in X_2$,

$$\langle T_2 x, J_{Y_2} T_2 x_2 \rangle_{Y_2} = \langle T_2 x, J_Y T_2 x_2 \rangle_Y$$

since $T_2 x \in Y_2$. Similarly $J_{X_2} z - J_X z \in X_2^0$ for all $z \in X_2$ and

$$\langle x, J_{X_2} x_2 \rangle_{X_2} = \langle x, J_X x_2 \rangle_X,$$

if $x \in X_2$. We therefore have

$$\langle T_2 x, J_Y T_2 x_2 \rangle_Y = \nu_2 \langle x, J_X x_2 \rangle_X, \quad x \in X_2.$$

Continuing in this way we obtain elements $x_1, x_2, \ldots, x_n$ of X, all with unit norm, subspaces $M_1, M_2, \ldots, M_n$ of X^* and $N_1, N_2, \ldots, N_n$ of Y^*, where for $k = 1, 2, \ldots, n$,

$$M_k = \text{sp}\{J_X x_1, \ldots, J_X x_k\}, \quad N_k = \text{sp}\{J_Y T x_1, \ldots, J_Y T x_k\}, \tag{2.2.8}$$

and decreasing families $X_1, \ldots, X_n$ and $Y_1, \ldots, Y_n$ of subspaces of X and Y respectively given by

$$X_k = {}^0 M_{k-1}, \quad Y_k = {}^0 N_{k-1}, \quad k = 2, \ldots, n. \tag{2.2.9}$$

Moreover, for each $k \in \{1, \ldots, n\}$, T maps X_k into Y_k, $x_k \in X_k$ and with $T_k := T \upharpoonright X_k$, $\lambda_k = \|T_k\|$, $\nu_k = \lambda_k \mu_Y(\lambda_k)$, we have

$$\langle T_k x, J_{Y_k} T_k x_k \rangle_{Y_k} = \nu_k \langle x, J_{X_k} x_k \rangle_{X_k} \tag{2.2.10}$$

for all $x \in X_k$ and so

$$T_k^* J_{Y_k} T_k x_k = \nu_k J_{X_k} x_k. \tag{2.2.11}$$

On identifying Y_k^* with Y^*/Y_k^0 and X_k^* with X^*/X_k^0, we see as before that (2.2.10) is equivalent to

$$\langle T_k x, J_Y T_k x_k \rangle_Y = \nu_k \langle x, J_X x_k \rangle_X. \tag{2.2.12}$$

Since $T x \in Y_k = {}^0 N_{k-1}$,

$$\langle T x_k, J_Y T x_l \rangle_Y = 0 \quad \text{if} \quad l < k \tag{2.2.13}$$

and by (2.2.12),
$$\langle x_k, J_X x_l \rangle_X = 0 \quad \text{if} \quad l < k. \tag{2.2.14}$$

In terms of the semi-inner product $(\cdot, \cdot)_X$ defined in Definition 1.1.29, (2.2.14) implies that
$$(x_l, x_k)_X = 0 \quad \text{if} \quad l < k. \tag{2.2.15}$$

The process stops with λ_n, x_n and X_{n+1} if and only if the restriction of T to X_{n+1} is the zero operator.

Proposition 2.2.4. *The restriction of T to X_{n+1} is the zero operator if and only if T is of rank n. If T is of infinite rank, then (λ_n) is infinite and converges to zero, and*
$$\ker(T) \supset \bigcap_{n \in \mathbb{N}} X_n. \tag{2.2.16}$$

Proof. For $x \in X$ and $k \geq 2$, put
$$w_k = x - \sum_{j=1}^{k-1} \xi_j x_j, \quad \xi_j = \xi_j(x), \tag{2.2.17}$$

where the ξ_j are chosen such that $w_k \in X_k$. We claim that this choice is possible and is uniquely given by
$$\xi_1 = \langle x, J_X x_1 \rangle_X, \quad \xi_l = \langle x - \sum_{j=1}^{l-1} \xi_j x_j, J_X x_l \rangle_X \quad \text{for} \quad 2 \leq l \leq k-1. \tag{2.2.18}$$

Since $\langle w_2, J_X x_1 \rangle_X = \langle x, J_X x_1 \rangle_X - \xi_1 \langle x_1, J_X x_1 \rangle_X = \langle x, J_X x_1 \rangle_X - \xi_1 = 0$, we have that $w_2 \in X_2$. Suppose that $w_{k-1} \in X_{k-1}$, and so $\langle w_{k-1}, J_X x_m \rangle_X = 0$ whenever $1 \leq m < k-1$. Then, for $1 \leq m < k-1$, it follows from $w_k = w_{k-1} - \xi_{k-1} x_{k-1}$ that
$$\langle w_k, J_X x_m \rangle_X = -\xi_{k-1} \langle x_{k-1}, J_X x_m \rangle_X = 0.$$

by (2.2.14). Moreover
$$\langle w_k, J_X x_{k-1} \rangle_X = \langle x - \sum_{j=1}^{k-2} \xi_j x_j, J_X x_{k-1} \rangle_X - \xi_{k-1} \langle x_{k-1}, J_X x_{k-1} \rangle_X = 0$$

by (2.2.18). Our claim that $w_k \in X_k$ for all k therefore follows by induction.

  If the restriction of T to X_{n+1} is the zero operator, then $T w_{n+1} = 0$ and hence $Tx = \sum_{j=1}^{n} \xi_j T x_j$. Therefore T is of finite rank and its range is the linear space spanned by $T x_1, \ldots, T x_n$.

  If T is not of finite rank, the process of constructing λ_n, x_n and X_{n+1} continues indefinitely. Since $T x_n \in {}^0 N_{n-1}$, (2.2.13) yields
$$\langle T x_n, \tilde{J}_Y T x_m \rangle_Y = 0 \quad \text{if} \quad m < n.$$

Thus if $m < n$,

$$\lim_{k\to\infty} \lambda_k \leq \|Tx_m\|_Y = \langle Tx_m, \tilde{J}_Y Tx_m\rangle_Y = \langle Tx_m - Tx_n, \tilde{J}_Y Tx_m\rangle_Y$$

$$\leq \|Tx_m - Tx_n\|_Y \|\tilde{J}_Y Tx_m\|_{Y^*} = \|Tx_m - Tx_n\|_Y.$$

Since (x_n) is bounded and T is compact, some subsequence of (Tx_n) must converge. Therefore (λ_k) must converge to zero.

If $x \in \bigcap_{n\in\mathbb{N}} X_n$, then for all $n \in \mathbb{N}, \|Tx\|_Y \leq \lambda_n \|x\|_X \to 0$ as $n \to \infty$. Thus $x \in \ker(T)$ and the proof is complete. $\qquad\square$

We summarise the previous results for the case when T is of infinite rank in the following theorem.

Theorem 2.2.5. *Suppose that T has infinite rank and that X and Y are reflexive with strictly convex duals. Then for each $k \in \mathbb{N}$, there exist $x_k \in X_k$ and $\lambda_k \in \mathbb{R}$ such that (2.2.11), (2.2.12), (2.2.13) and (2.2.14) are satisfied, where $\nu_k = \lambda_k \mu_Y(\lambda_k)$, and $\lim_{k\to\infty} \lambda_k = 0$. Also (2.2.16) holds.*

We shall in fact prove in Proposition 2.2.10 below that if T is of infinite rank, then $\ker(T) = \bigcap_{n\in\mathbb{N}} X_n$.

It is illuminating to recast the above iterative procedure in terms of the notion of orthogonality investigated by James in [52] (see Definition 1.1.30 and Proposition 1.1.31), which gives the analysis a more geometric flavour. In the language of j-orthogonality, (2.2.9) yields

$$\text{sp}\{x_i\} \perp^j X_{k+1}, \quad \text{sp}\{y_i\} \perp^j Y_{k+1}, \quad \text{for all} \quad i \leq k, \tag{2.2.19}$$

where $y_j = \lambda_j^{-1} Tx_j$.

We shall call the sequences (x_k) and (λ_k) the **j-eigenvectors** and **j-eigenvalues** of T respectively. In Remark 2.2.8 below, we shall see that when X and Y are Hilbert spaces, $T^*Tx_k = \lambda_k^2 x_k$ so that the λ_k are singular values of T. However, in general, apart from the coincidence of the first j-eigenvectors and j-eigenvalues with the first classical eigenvectors and eigenvalues, it seems unlikely that any close connections exist between the $(x_k), (\lambda_k)$ and the eigenvectors and eigenvalues of T. Numerical evidence for this is discussed in [38], section 5.

We now show that the coefficients ξ_k of (2.2.18) are given by an explicit formula. Setting $\hat{\xi} := (\xi_1, \xi_2, \ldots, \xi_n)^t$, where the superscript t denotes the transpose, we observe that

$$\Gamma(x_1, x_2, \ldots, x_n)\hat{\xi} = (a_1, a_2, \ldots, a_n)^t, \tag{2.2.20}$$

where in terms of the semi-inner product $(\cdot, \cdot)_X$ of Definition 1.1.29, $a_j = (x_j, x)_X$, $j = 1, 2, \ldots, n$ and with $a_{j,k} := (x_k, x_j)_X, \Gamma$ is the lower-triangular $n \times n$ matrix with kth row

$$(a_{k,1}, a_{k,2}, \ldots, a_{k,k-1}, 1, 0, \ldots, 0)$$

if $k \geq 2$, and first row $1, 0, \ldots, 0$. The matrix Γ is invertible, its inverse Γ^{-1} being also lower-triangular. Let the kth row of Γ^{-1} be

$$(A_{k,1}, A_{k,2}, \ldots, A_{k,k-1}, 1, 0, \ldots, 0)$$

and hence its jth column is

$$(0, \ldots, 0, 1, A_{j+1,j}, A_{j+2,j}, \ldots, A_{n,j})^t .$$

Then for $j \leq k - 2$,

$$a_{k,j} + \sum_{t=1}^{k-j-1} a_{k,j+t} A_{j+t,j} + A_{k,j} = 0. \tag{2.2.21}$$

and when $j = k - 1$,

$$a_{k,k-1} + A_{k,k-1} = 0. \tag{2.2.22}$$

These give

$$\xi_k = a_k + \sum_{i=1}^{k-1} A_{k,i} a_i. \tag{2.2.23}$$

We shall now prove that

$$A_{k,,j} = \sum (-1)^m \prod_{i=1}^{m} a_{l_i, l_{i+1}} \quad (j < k), \tag{2.2.24}$$

where the sum is over all $m \in \{1, \ldots, k - j\}$ and all sequences of natural numbers $(l_i)_{i=1}^{m+1}$ such that $k = l_1 > \cdots > l_{m+1} = j$.

It is routine to check that (2.2.24) holds for any k when $j = k - 1$ or $k - 2$. Suppose that for some k we have, whenever $j + 1 \leq s \leq k - 1$,

$$A_{s,,j} = \sum (-1)^m \prod_{i=1}^{m} a_{l_i, l_{i+1}}, \tag{2.2.25}$$

where the sum is over all $m \in \{1, \ldots, s - j\}$ and all sequences of natural numbers $(l_i)_{i=1}^{m+1}$ such that $s = l_1 > \cdots > l_{m+1} = j$. We must prove that (2.2.25) holds with $s = k$; that is, we want to prove that

$$A_{k,,j} = \sum (-1)^m \prod_{i=1}^{m} a_{l_i, l_{i+1}}, \tag{2.2.26}$$

where the sum is over all $m \in \{1, \ldots, k - j\}$ and all sequences of natural numbers $(l_i)_{i=1}^{m+1}$ such that $k = l_1 > \cdots > l_{m+1} = j$. From (2.2.21) and the inductive hypothesis,

$$A_{k,j} = -a_{k,j} - \sum_{t=1}^{k-j-1} a_{k,j+t} \sum (-1)^m \prod_{i=1}^{m} a_{l_i, l_{i+1}}, \tag{2.2.27}$$

where for each $t \in \{1, \ldots, k-j-1\}$ we have $m \in \{1, \ldots, t\}$ and

$$j + t = l_1 > \cdots > l_{m+1} = j \qquad (2.2.28)$$

in the inner sum. It is now a question of showing that (2.2.26) and (2.2.27) coincide. The contribution to (2.2.26) arising from $m = 1$ gives $-a_{k,j}$, which is present in (2.2.27). Now we compare the coefficients of $a_{k,j+t}$ $(1 \le t \le k-j-1)$ in both expressions. In (2.2.27) this is

$$-\sum (-1)^m \prod_{i=1}^{m} a_{l_i, l_{i+1}}, \qquad (2.2.29)$$

where $m \in \{1, \ldots, t\}$ and $j + t = l_1 > \cdots > l_{m+1} = j$. In (2.2.26) the term involving $a_{k,j+t}$ is

$$-a_{k,j+t} \sum (-1)^{m-1} \prod_{i=2}^{m} a_{l_i, l_{i+1}},$$

where the sum is over all l_i such that $l_2 = j + t > \cdots > l_{m+1} = j$. Hence the coefficient of $a_{k,j+t}$ is

$$-\sum (-1)^{m-1} \prod_{i=2}^{m} a_{l_i, l_{i+1}} = -\sum (-1)^s \prod_{i=1}^{s} a_{l_i, l_{i+1}},$$

where the sum is over all l_i such that for some s, $l_1 = j + t > \cdots > l_s = j$. This coincides with (2.2.29), and thus (2.2.26) holds. Completion of this inductive step means that (2.2.24) is established. The formula for ξ_k given by (2.2.23) and (2.2.24) is in stark contrast to that when X is a Hilbert space, for then $a_{j,k} = \delta_{j,k}$ and $\xi_k = (x_k, x)_X$.

2.2.2 The linear projections S_k

We remind the reader that it is supposed that X and Y are real reflexive Banach spaces with strictly convex duals. A key role is hereafter played by the family of maps

$$S_k : X \to Z_{k-1} := \mathrm{sp}\{x_1, \ldots, x_{k-1}\}, \quad k \ge 2,$$

defined by the condition that $x - S_k x \in X_k$ for all $x \in X$. As we saw in the proof of Proposition 2.2.4, these maps are uniquely given by

$$S_k x := \sum_{j=1}^{k-1} \xi_j(x) x_j, \qquad (2.2.30)$$

where the coefficients $\xi_j(x)$ satisfy (2.2.18). Hence S_k is linear, and since

$$x - S_k^2 x = (x - S_k x) - S_k(S_k x - x) \in X_k,$$

it follows from the uniqueness that $S_k^2 = S_k$. Thus S_k is a linear projection of X onto Z_{k-1}. In fact we have

Lemma 2.2.6. *The spaces X and X^* have the direct sum decompositions*

$$X = X_k \oplus Z_{k-1}, \quad X^* = M_{k-1} \oplus Z_{k-1}^0 \qquad (2.2.31)$$

for each $k \geq 2$. The operators S_k, S_k^* are respectively linear projections of X onto Z_{k-1} and X^* onto M_{k-1}. Furthermore, for all $k \in \mathbb{N}$,

$$X_k = X_{k+1} \boxplus \operatorname{sp}\{x_k\}. \qquad (2.2.32)$$

Proof. The decomposition for X follows from $I = (I - S_k) + S_k$, where I is the identity map of X to itself, since $I - S_k$ maps X into X_k by definition, and S_k has range Z_{k-1}. It is unique in view of (2.2.19).

From (2.2.30) and (2.2.18), we have

$$S_k x = S_{k-1} x + \langle \{x - S_{k-1} x\}, J_X x_{k-1} \rangle_X x_{k-1}$$

and so, on setting $E_j := \langle \cdot, J_X x_j \rangle_X x_j$,

$$S_k = S_{k-1} + E_{k-1}(I - S_{k-1}),$$

which yields

$$I - S_k = (I - E_{k-1}) \cdots (I - E_1), \quad k \geq 2. \qquad (2.2.33)$$

For $x \in X$ and $x^* \in X^*$,

$$\langle E_{k-1} x, x^* \rangle_X = \langle x, \langle x_k, x^* \rangle_X J_X x_k \rangle_X.$$

Thus the map $E_{k-1} : X \to X$ has adjoint $E_{k-1}^* : X^* \to X^*$ given by $E_{k-1}^* = \langle x_k, \cdot \rangle_X J_X x_k$, and by (2.2.33),

$$I^* - S_k^* = (I^* - E_1^*) \cdots (I^* - E_{k-1}^*). \qquad (2.2.34)$$

It readily follows by induction that S_k^* and $I^* - S_k^*$ have ranges M_{k-1} and Z_{k-1}^0, respectively, and hence $X^* = M_{k-1} \oplus Z_{k-1}^0$. Also the identity $(S_{k-1}^*)^2 = S_{k-1}^*$ is an easy consequence of $S_{k-1}^2 = S_{k-1}$.

From (2.2.31), $S_k x = 0$ for $x \in X_k$. Since $S_{k+1} x = S_k x + \xi_k(x) x_k$ it follows that for all $x \in X_k$, $S_{k+1} x = \xi_k(x) x_k$, and as $(I - S_{k+1}) x \in X_{k+1} \subset X_k$, we therefore have $X_k = X_{k+1} \oplus \operatorname{sp}\{x_k\}$; thus (2.2.32) is implied by (2.2.19). $\qquad \square$

In view of Lemma 2.2.6, we can write the identity (2.2.12) as

$$\langle T(I - S_k)x, J_Y T x_k \rangle_Y = \nu_k \langle (I - S_k)x, J_X x_k \rangle_X, \quad \text{for all} \quad x \in X. \qquad (2.2.35)$$

Therefore

$$(T^* J_Y T - \nu_k J_X) x_k \in X_k^0 = M_{k-1} \qquad (2.2.36)$$

and

$$(I^* - S_k^*)(T^* J_Y T - \nu_k J_X) x_k = 0. \qquad (2.2.37)$$

Remark 2.2.7. Suppose that T is of infinite rank and that X and Y are reflexive with strictly convex duals. If λ is a non-zero eigenvalue of T corresponding to a normalised eigenvector x, then

$$\langle x, \lambda^{-1}T^*\tilde{J}_X x\rangle_X = \langle \lambda^{-1}Tx, \tilde{J}_X x\rangle_X = \langle x, \tilde{J}_X x\rangle_X = 1,$$

so that by the strict convexity and reflexivity of X^*

$$T^*\tilde{J}_X x = \lambda \tilde{J}_X x.$$

This means that $\tilde{J}_X x$ is an eigenvector of T^* corresponding to the eigenvalue λ. Moreover, since $\tilde{J}_X(\lambda x) = (\operatorname{sgn}\lambda)\,\tilde{J}_X x$,

$$T^*\tilde{J}_X Tx = T^*\tilde{J}_X(\lambda x) = (\operatorname{sgn}\lambda)T^*\tilde{J}_X x = |\lambda|\tilde{J}_X x.$$

Thus $|\lambda|$ is a j-eigenvalue of T with corresponding j-eigenvector x.

Remark 2.2.8. When X and Y are Hilbert spaces, the duals are identified with the original spaces and the supporting functionals $\tilde{J}_X, \tilde{J}_Y$ become the identity maps. We then choose the gauge functions $\mu_X(t) = \mu_Y(t) = t$. Also $S_k^* = S_k$ and the direct sums in Lemma 2.2.6 are now orthogonal sums since $Z_{k-1} = M_{k-1}$. By (2.2.13), $T^*Tx_k \in X_k$ and so $S_k(T^*T - \nu_k I)x_k = 0$, where now $\nu_k = \lambda_k^2$. We therefore infer from (2.2.37) that $T^*Tx_k = \lambda_k^2 x_k$ and so λ_k is a singular value of T.

Lemma 2.2.9. *For all $j, k \in \mathbb{N}$,*

$$\xi_j(x_k) = \delta_{j,k}, \tag{2.2.38}$$

where $\delta_{j,k}$ is the Kronecker delta.

Proof. Let $k \in \mathbb{N}$ be fixed and choose any $n > k$. Then $x_k = \sum_{j=1}^{n-1}\delta_{j,k}x_j$ and $x_k - \sum_{j=1}^{n-1}\xi_j(x_k)x_j = x_k - S_n x_k \in X_n$. Hence

$$\sum_{j=1}^{n-1}(\delta_{j,k} - \xi_j(x_k))x_j \in X_n.$$

This implies that $\xi_j(x_k) = \delta_{j,k}$ for all $j \leq n-1$ and the asserted result follows. A direct inductive proof may also be given. $\qquad\square$

2.2.3 The nonlinear projections $P_k : X \to X_k$

Suppose that X and X^* are strictly convex. The projection P_k of X onto X_k is in general non-linear; $z_k = P_k x$ is the nearest point in X_k to $x \in X$, this being uniquely defined since X is assumed to be strictly convex. As $\|z_k - x\|_X \leq \|x\|_X$ it follows that

$$\|z_k\|_X \leq 2\|x\|_X \tag{2.2.39}$$

and so

$$\|Tz_k\|_Y \leq \lambda_k \|z_k\|_X \to 0 \qquad (2.2.40)$$

as $k \to \infty$. Since

$$\|x - z_k\|_X = \inf\{\|x - z_k + ty\|_X : y \in X_k\},$$

we see that for all $y \in X_k$,

$$\frac{d}{dt}\|x - z_k + ty\|_X\bigg|_{t=0} = 0,$$

and hence

$$\langle y, \tilde{J}_X(x - z_k)\rangle_X = 0.$$

This in turn implies that $\langle y, J_X(x - z_k)\rangle_X = 0$. Consequently $J_X(x - z_k) \in M_{k-1}$ and so

$$(I - P_k)x \in J_X^{-1}M_{k-1}. \qquad (2.2.41)$$

Thus any $x \in X$ can be written as $x = P_k x + (I - P_k)x$ where $P_k x \in X_k$ and $(I - P_k)x \in J_X^{-1}M_{k-1}$. If $x \in X_k \cap J_X^{-1}M_{k-1}$ then since $X_k =^0 M_{k-1}$ it follows that

$$\langle x, J_X x\rangle_X = 0$$

and hence $x = 0$. Furthermore, for any $x \in X_k$ and $y \in J_X^{-1}M_{k-1}, y \neq 0$,

$$(y, x)_X := \|y\|_X\langle x, \tilde{J}_X y\rangle_X = \frac{\|y\|_X}{\mu_X(\|y\|_X)}\langle x, J_X y\rangle_X = 0,$$

this being trivially true when $y = 0$. Hence $J_X^{-1}M_{k-1} \perp^j X_k$. It therefore follows that for any $k \in \mathbb{N}$,

$$X = X_k \biguplus J_X^{-1}M_{k-1}. \qquad (2.2.42)$$

Note that (2.2.42) follows from Alber's theorem, Theorem 1.1.33, when X is uniformly convex and uniformly smooth.

By Lemma 2.2.6,

$$J_X(x - P_k x) = S_k^* J_X(x - P_k x)$$

and so

$$(x - P_k x) = J_X^{-1}S_k^* J_X(x - P_k x). \qquad (2.2.43)$$

On using (2.2.40) we have that $\|TP_k x\|_Y \to 0$ as $k \to \infty$ and hence

$$Tx = \lim_{k\to\infty} TJ_X^{-1}S_k^* J_X(x - P_k x). \qquad (2.2.44)$$

Since $\bigcap_{k\in\mathbb{N}} X_n \subset \ker(T)$ by Proposition 2.2.4, it follows that if T is of infinite rank, there is a strictly increasing sequence $(k(j))_{j\in\mathbb{N}}$ of natural numbers such that the weak limit $w - \lim_{j\to\infty} z_{k(j)}$ exists and lies in $\ker(T)$. Thus if $\ker(T) = \{0\}$, we have from (2.2.43)

$$x = w - \lim_{j\to\infty} J_X^{-1} S_k^* J_X(x - P_{k(j)}x) \tag{2.2.45}$$

for all $x \in X$. If X is a Hilbert space, $P_{k(j)} = I - S_{k(j)}$, $(S_{k(j)}^*)^2 = S_{k(j)}^* = S_{k(j)}$ in (2.2.45) and so $x = w - \lim_{j\to\infty} S_{k(j)}x$. This implies that x is the strong limit of some linear combination of the x_i, by Mazur's theorem; see [65], Proposition 1.3.1. We may therefore write

$$x = \sum_{i=1}^{\infty} \eta_i(x)x_i,$$

in the sense of strong convergence, where $\eta_i(x) = (x, x_i)_X$. In other words, the result that (x_i) is a basis of the Hilbert space X when T is of infinite rank and $\ker(T) = \{0\}$, established in Section 2.1, is recovered. Furthermore

$$Tx = \sum_{i=1}^{\infty} \eta_i(x)Tx_i.$$

Although the projections $P_k : X \to X_k$ are nonlinear in general, the next lemma (established in [41]) shows that they are endowed with certain linearity properties in view of a relationship between them and the linear projections S_n.

Lemma 2.2.10. *For all $x \in X$ and $n \in \mathbb{N} \setminus \{1\}$,*

$$x - S_n x = P_n P_{n-1} \cdots P_2 x. \tag{2.2.46}$$

If X is a Hilbert space $I - S_n = P_n$.

Proof. We prove the lemma by induction. Since $P_2 x$ is the nearest point in X_2 to x, it follows that for all $\alpha \in \mathbb{R}$ and all $z \in X_2$,

$$\|x - P_2 x\|_X \le \|x - P_2 x + \alpha z\|_X,$$

which implies that $x - P_2 x \perp^j X_2$; see Definition 1.1.30 and Proposition 1.1.31. By (2.2.32) with $k = 1$ (and $X_1 = X$), bearing in mind that the decomposition is unique, it follows that $x - P_2 x \in \mathrm{sp}\{x_1\}$ and so since S_2 is the unique map with the properties that S_2 and $(I - S_2)$ have ranges $\mathrm{sp}\{x_1\}$, X_2, respectively, we have that $(I - P_2) = S_2$.

Suppose that $x - S_n x = P_n P_{n-1} \cdots P_2 x$ holds for some $n \in \mathbb{N}$ and all $x \in X$; to simplify notation set $H_n = P_n P_{n-1} \cdots P_2$. Then, for all $x \in X$, $z \in X_{n+1}$ and $\alpha \in \mathbb{R}$,

$$\|H_n x - P_{n+1} H_n x\|_X \le \|H_n x - P_{n+1} H_n x + \alpha z\|_X.$$

Hence $H_n x - P_{n+1} H_n x \perp^j X_{n+1}$, and since $X_{n+1} \subset X_n$, we have that $H_n x - P_{n+1} H_n x \in X_n$. Therefore (2.2.32) implies that $H_n x - P_{n+1} H_n x \in \mathrm{sp}\{x_n\} \subset Z_n$ and consequently, in view of Lemma 2.2.6,

$$H_n x - P_{n+1} H_n x = S_{n+1}(H_n x - P_{n+1} H_n x) = S_{n+1} H_n x.$$

It follows from the induction hypothesis that

$$x - S_n x - P_{n+1} H_n x = S_{n+1}(x - S_n x) = S_{n+1} x - S_n x$$

and thus $x - S_{n+1} x = P_{n+1} H_n x$. This completes the proof of (2.2.46).

If X is a Hilbert space, we have from Lemma 2.2.6 that it has the orthogonal sum decomposition $X = X_n \oplus Z_{n-1}$ and $I - S_n$ is the projection onto X_n. Thus, for all $x \in X$ and $y \in X_n$,

$$\|x - y\|_X^2 = \|(I - S_n)x - y\|_X^2 + \|S_n x\|_X^2 \geq \|S_n x\|_X^2 = \|x - (I - S_n)x\|_X^2,$$

and this implies that $I - S_n = P_n$. $\square$

2.2.4   The main convergence theorems

To fulfill our objective of deriving convergence results in a Banach space setting, we need the following abstract result.

Lemma 2.2.11. *Let $\mathcal{L}(X)$ be the set of all closed linear subspaces of a Banach space X, let $\mathbb{S} \subset \mathcal{L}(X)$ and put $L = \bigcap_{S \in \mathbb{S}} S$, $N = \overline{\bigcup_{S \in \mathbb{S}} S^0}$. Then $L^0 = N$ and for all $x \in X$,*

$$\|x\|_{X/L} = \sup_{S \in \mathbb{S}} \|x\|_{X/S}. \tag{2.2.47}$$

Proof. Let $\Phi : \mathcal{L}(X) \to \mathcal{L}(X^*)$ be given by $\Phi(Z) = Z^0$ $(Z \in \mathcal{L}(X))$. We recall that given any linear subspace G of a Banach space, its polar G^0 is closed and $\overline{G} =^0 (G^0)$; a corresponding result holds for subspaces of the dual space. Hence Φ is bijective; it also reverses inclusion. Thus for any $S \in \mathbb{S}$, $L \subset S$ and so $S^0 \subset L^0$; hence $\bigcup_{S \in \mathbb{S}} S^0 \subset L^0$ and $N \subset L^0$. Moreover, for all $S \in \mathbb{S}$, $S^0 \subset N$, which shows that $\Phi^{-1}(N) \subset \bigcap_{S \in \mathbb{S}} S = L$, whence $L^0 \subset N$. Consequently $L^0 = N$. This implies that $(X/L)^*$ is isometrically isomorphic to N, and

$$\sup\{|\langle x, y \rangle_X| : y \in N, \|y\|_{X^*} \leq 1\} = \sup\{|\langle x, y \rangle_X| : y \in \bigcup_{S \in \mathbb{S}} S^0, \|y\|_{X^*} \leq 1\}$$

$$= \sup\{|\langle x, y \rangle_X| : y \in S^0, \|y\|_{X^*} \leq 1\}.$$

The result follows. $\square$

We now apply Lemma 2.2.11 with $L = X_\infty = \bigcap_{k \in \mathbb{N}} X_k$. Since $X_k^0 = (^0 M_{k-1})^0 = M_{k-1}$, we have that $N = \overline{\bigcup_{k \in \mathbb{N}} M_k}$ and hence by the lemma,

$$X_\infty^0 = \overline{\bigcup_{k \in \mathbb{N}} M_k} \tag{2.2.48}$$

and

$$\|x\|_{X/X_\infty} = \sup_{k\in\mathbb{N}} \|x\|_{X/X_k} = \lim_{k\to\infty} \|x\|_{X/X_k}, \qquad (2.2.49)$$

the last equality in (2.2.49) being a consequence of the fact that the subspaces X_k decrease with k and hence the norms on X/X_k increase. Note that we have omitted the canonical maps of X into X/X_∞ and X into X/X_k to simplify notation, and we shall continue with this abuse of notation whenever the meaning is obvious. However, when their presence makes the argument clearer they are retained.

Lemma 2.2.12. *Suppose that X is reflexive and that X and X^* are strictly convex. Put $X_\infty = \cap_{k\in\mathbb{N}} X_k$ and let $P_k : X \to X_k$, $P_\infty : X \to X_\infty$ be the projections discussed in Section 2.2.3. Then for all $x \in X$, $P_k x \rightharpoonup P_\infty x$ and $\|x - P_k x\|_X \to \|x - P_\infty x\|_X$ as $k \to \infty$. If X is assumed to be uniformly convex, then $P_k x \to P_\infty x$ as $k \to \infty$.*

Proof. Since

$$\|x - P_k x\|_X = \|x\|_{X/X_k} \le \|x\|_X,$$

it follows that $\|P_k x\|_X \le 2\|x\|_X$; hence $\{P_k x\}$ has a subsequence that converges weakly, to $y \in X_\infty$, say. We claim that $y = P_\infty x$. For if not, then

$$\|x - y\|_X > \|x - P_\infty x\|_X = \|x\|_{X/X_\infty}.$$

Thus

$$\|x - P_k x\|_X \ge \left\langle x - P_k x, \widetilde{J_X}(x - y) \right\rangle_X \to \left\langle x - y, \widetilde{J_X}(x - y) \right\rangle_X$$
$$= \|x - y\|_X > \|x\|_{X/X_\infty},$$

and this implies that for some $k \in \mathbb{N}$,

$$\|x - P_k x\|_X > \|x\|_{X/X_\infty}.$$

But this means that

$$\|x\|_{X/X_k} > \|x\|_{X/X_\infty},$$

which contradicts the fact that $X_\infty \subset X_k$. Thus every weakly convergent subsequence of $\{P_k x\}$ has weak limit $P_\infty x$, from which it follows by a standard contradiction argument that the whole sequence $\{P_k x\}$ converges weakly to $P_\infty x$. However, by (2.2.49),

$$\|x - P_\infty x\|_X = \|x\|_{X/X_\infty} = \lim_{k\to\infty} \|x\|_{X/X_k} = \lim_{k\to\infty} \|x - P_k x\|_X.$$

By Proposition 1.1.13, it follows that $P_k x \to P_\infty x$ if X is uniformly convex. □

Lemma 2.2.13. *Let S_k, $k \ge 2$, be the linear projections of X onto $Z_{k-1} = \mathrm{sp}\{x_1, \ldots, x_{k-1}\}$ given by (2.2.30), where the coefficients ξ_j satisfy (2.2.18), and let P_k be the projection of X onto X_k. Then $(I - P_k)S_k x \to x$ in X/X_∞.*

Proof. From $(I - P_k)S_k x - x = (S_k x - x) - P_k S_k x \in X_k \subset X_n =^0 M_{n-1}$ for $k \geq n$, it follows that for $y \in \bigcup_{n \geq 1} M_n$,

$$\langle (I - P_k)S_k x - x, y \rangle_X \to 0 \tag{2.2.50}$$

as $k \to \infty$. Furthermore, since $S_k x - x \in X_k$ for all $x \in X$, we have

$$\|S_k x - P_k S_k x\|_X = \|S_k x\|_{X/X_k} = \|x\|_{X/X_k} \leq \|x\|_{X/X_\infty} \leq \|x\|_X \tag{2.2.51}$$

and so $S_k x - P_k S_k x - x$ is bounded on X.

Let $\phi : X \to X/X_\infty$ be the canonical map. By Proposition 2.2.1, its adjoint ϕ^* is an isometric isomorphism of $(X/X_\infty)^*$ onto X_∞^0. We infer from (2.2.50), on noting (2.2.48), that for all $y \in \bigcup_{n \in \mathbb{N}} M_n = X_\infty^0$,

$$\langle \phi[(I - P_k)S_k x - x], (\phi^*)^{-1} y \rangle_{X/X_\infty} = \langle (I - P_k)S_k x - x, y \rangle_X \to 0.$$

The result follows since $(\phi^*)^{-1}(X_\infty^0) = (X/X_\infty)^*$. $\qquad\square$

Theorem 2.2.14. *Suppose that X is reflexive and that X and X^* are strictly convex. Then for all $x \in X$,*

$$\lim_{n \to \infty} \|S_n x - P_n S_n x\|_{X/X_\infty} = \|x\|_{X/X_\infty} .$$

If X is uniformly convex, then

$$x = \lim_{n \to \infty} [S_n x - P_n S_n x] \quad in \ X/X_\infty$$

and

$$x = \lim_{n \to \infty} [S_n x - P_n S_n x] + P_\infty x \quad in \ X.$$

Proof. If $\|y\|_{X^*} \leq 1$, then

$$\left| \langle \phi_\infty (S_n x - P_n S_n x), (\phi_\infty^*)^{-1} y \rangle_{X/X_\infty} \right|$$
$$= |\langle S_n x - P_n S_n x, y \rangle_X | \leq \|S_n x - P_n S_n x\|_X$$
$$= \|S_n x\|_{X/X_n} = \|x\|_{X/X_n} \leq \|x\|_{X/X_\infty} .$$

Thus

$$\|x\|_{X/X_n} = \|S_n x - P_n S_n x\|_{X/X_n} \leq \|S_n x - P_n S_n x\|_{X/X_\infty} \leq \|x\|_{X/X_\infty} ,$$

by (2.2.51), and the first part follows from (2.2.49). If X is uniformly convex, the claimed result in X/X_∞ follows from Lemma 2.2.13.

Since $S_n x - x \in X_n$, we also have

$$(I - P_n)S_n x - (I - P_n)x = (S_n x - x) - P_n(S_n x - x)$$
$$= (S_n x - x) - (S_n x - x) = 0.$$

Hence,

$$x = (I - P_n)S_n x + P_n x$$

and the final result follows from Lemma 2.2.12 $\qquad\square$

Theorem 2.2.15. *Suppose that X is reflexive, X and X^* are strictly convex, and* rank $T = \infty$. *Then for all $x \in X$,*

$$Tx = \lim_{n \to \infty} \left\{ \sum_{i=1}^{n-1} \lambda_i \xi_i(x) y_i - T P_n S_n x \right\}, \quad y_i = Tx_i / \|Tx_i\|_Y = Tx_i / \lambda_i.$$

Proof. Since $x - S_n x + P_n S_n x \in X_n$ and $X_\infty \subset X_n$, we have that

$$\|T(x - S_n x + P_n S_n x)\|_Y \leq \lambda_n \left(\|x\|_X + \|S_n x - P_n S_n x\|_X \right)$$
$$= \lambda_n \left(\|x\|_X + \|\phi_n(S_n x)\|_{X/X_n} \right)$$
$$= \lambda_n \left(\|x\|_X + \|\phi_n x\|_{X/X_n} \right)$$
$$\leq \lambda_n \left(\|x\|_X + \|\phi_\infty x\|_{X/X_\infty} \right) \to 0. \qquad \square$$

Note that as $X_l \subset X_k$ for $k \leq l$, (2.2.12) gives

$$\langle Tx_l, J_Y Tx_k \rangle_Y = \nu_k \langle x_l, J_X x_k \rangle_X.$$

Moreover, with $\mu_Y(1) = 1$, $J_Y Tx_k = \mu_Y(\lambda_k) J_Y y_k$. Therefore, in terms of the semi-inner product in Definition 1.1.29 we have, for $k \leq l$, that

$$(y_k, y_l)_Y = \langle y_l, J_Y y_k \rangle_Y = \frac{1}{\lambda_l \mu_Y(\lambda_k)} \langle Tx_l, J_Y Tx_k \rangle_Y$$
$$= \frac{\lambda_k}{\lambda_l} (x_k, x_l)_X = \delta_{k,l} \quad (k \leq l), \tag{2.2.52}$$

by (2.2.15) and since $\|x_k\|_X = 1$.

Proposition 2.2.16. *Let X and X^* be strictly convex. If* rank T *is infinite, then* $\ker(T) = X_\infty$.

Proof. Suppose that $x \in X \backslash X_\infty$, so that $x \notin X_k$ for some k and, in view of our decomposition Lemma 2.2.6, $S_k x = \sum_{i=1}^{k-1} \xi_i(x) x_i \neq 0$. Let $\xi_l(x)$ be the first non-zero coefficient in this sum. By Theorem 2.2.15,

$$Tx = \lim_{n \to \infty} \left\{ \sum_{i=1}^{n-1} \lambda_i \xi_i(x) y_i - T P_n S_n x \right\}.$$

If $Tx = 0$, then

$$0 = \lim_{n \to \infty} \left\langle \sum_{i=1}^{n-1} \lambda_i \xi_i(x) y_i - T P_n S_n x, \tilde{J}_Y y_l \right\rangle_Y = \lambda_l \xi_l(x),$$

since $T P_n S_n x \in Y_n \subset Y_l$ for $n > l$ implies that $\langle T P_n S_n x, \tilde{J}_Y y_l \rangle_Y = 0$, and $\langle y_i, J_Y y_l \rangle_Y = 0$ for $i > l$ by (2.2.52). This contradiction shows that $\ker T \subset X_\infty$, and since we already know that the reverse inclusion holds by (2.2.16), the result follows. $\qquad \square$

Remark 2.2.17. Suppose that (x_k) is an infinite sequence of elements of the unit sphere of X with the semi-orthogonality property (2.2.15), namely, $(x_l, x_k)_X = 0$ for all $l, k \in \mathbb{N}$ with $l < k$. From this we can construct the subspaces X_k just as in (2.2.9), and it can be checked that the representation of elements x of X given in, for example, Theorem 2.2.14, still holds. In other words, the operator T is needed, so far as this particular representation is concerned, simply to establish the existence of a sequence $\{x_k\}$ with the appropriate properties.

Remark 2.2.17 has the following implication.

Define maps $R_k : Y \to \mathrm{sp}\{y_1, \ldots, y_{k-1}\}, k \geq 2$, to be the analogues of the maps S_k, determined by the conditions $y - R_k y \in Y_k$ for all $y \in Y$. As in the case of S_k, they are linear projections and are uniquely given by

$$R_k y = \sum_{j=1}^{k-1} \gamma_j(y) y_j, \tag{2.2.53}$$

where, for $j \geq 2$,

$$\gamma_j(y) = \left\langle y - \sum_{i=1}^{j-1} \gamma_i(y) y_i, J_Y y_j \right\rangle_Y, \quad \gamma_1(y) = \langle y, J_Y y_1 \rangle_Y. \tag{2.2.54}$$

Moreover, with $V_k := \mathrm{sp}\{y_1, y_2, \ldots, y_k\}$, we have an analogue of Lemma 2.2.6, namely,

Lemma 2.2.18. *The spaces Y and Y^* have the direct sum decompositions*

$$Y = Y_k \oplus V_{k-1}, \quad Y^* = N_{k-1} \oplus V_{k-1}^0 \tag{2.2.55}$$

for each $k \geq 2$. The operators R_k, R_k^* are respectively linear projections of Y onto V_{k-1} and Y^* onto N_{k-1}. Furthermore, for all $k \in \mathbb{N}$,

$$Y_k = Y_{k+1} \uplus \mathrm{sp}\{y_k\}. \tag{2.2.56}$$

Since T maps X_k into Y_k and Z_{k-1} into V_{k-1}, it follows that

$$R_k T x = T S_k x, \quad \text{for all } x \in X. \tag{2.2.57}$$

Furthermore, $Y_k \cap TX \subset TX_k$ by (2.2.12), and since we have already observed in the discussion following Proposition 2.2.3 that T maps X_k into Y_k, we have that

$$Y_k \cap TX = TX_k. \tag{2.2.58}$$

This implies that

$$\left(\bigcap_{n \in \mathbb{N}} Y_n \right) \cap TX = \{0\}. \tag{2.2.59}$$

For if $y \in \left(\bigcap_{n \in \mathbb{N}} Y_n\right) \cap TX$, then $y = Tx \in Y_n$ for all $n \in \mathbb{N}$ and so by (2.2.58) and Proposition 2.2.16, $x \in \bigcap_{n \in \mathbb{N}} X_n = \ker T$. Thus $y = 0$ and (2.2.59) follows.

If T has dense range in Y, that is, $\overline{TX} = Y$, then $(TX)^0 = \{0\}$ and we have from (2.2.59),

$$Y^* = \{0\}^0 = \left\{\left(\bigcap_{n \in \mathbb{N}} Y_n\right) \cap TX\right\}^0 = \overline{\left(\bigcap_{n \in \mathbb{N}} Y_n\right)^0 \cup (TX)^0}$$

$$= \overline{\bigcup_{n \in \mathbb{N}} Y_n^0} = \left(\bigcap_{n \in \mathbb{N}} Y_n\right)^0,$$

and hence

$$\bigcap_{n \in \mathbb{N}} Y_n = \{0\}. \tag{2.2.60}$$

These arguments now easily lead to

Theorem 2.2.19. *Suppose that Y is uniformly convex and Y^* is strictly convex and write Q_k for the (generally nonlinear) projections of Y onto Y_k. If T has dense range in Y, then for all $x \in X$, $Q_k Tx \to 0$ as $k \to \infty$ and*

$$Tx = \lim_{k \to \infty} (I - Q_k) T S_k x, \tag{2.2.61}$$

where for $k > 1$, $T S_k x = \sum_{j=1}^{k-1} \xi_j(x) T x_j = \sum_{j=1}^{k-1} \lambda_j \xi_j(x) y_j$.

Corollary 2.2.20. *Suppose that X^* is strictly convex and Y is a Hilbert space. If T has dense range in Y, then for all $x \in X$,*

$$Tx = \sum_{j=1}^{\infty} \lambda_j \xi_j(x) y_j, \tag{2.2.62}$$

where $\xi_j(x) = \lambda_j^{-1}(Tx, y_j)_Y$.

Proof. Since Y is a Hilbert space, $\{y_j\}$ is an orthonormal sequence in Y, $Y_k = N_{k-1}^{\perp}$ and N_{k-1} is the linear span of $\{Tx_1, \ldots, Tx_{k-1}\}$, which is the linear span of $\{y_1, \ldots, y_{k-1}\}$. Thus since $T S_k x \in N_{k-1}$, we have that $Q_k T S_k x = 0$. Therefore (2.2.62) follows from Theorem 2.2.19. $\square$

We also have the following counterpart of Lemma 2.2.10:

Lemma 2.2.21. *For all $y \in Y$ and $n \in \mathbb{N} \setminus \{1\}$,*

$$y - R_n y = Q_n Q_{n-1} \cdots Q_2 y. \tag{2.2.63}$$

2.2.5   A basis for X

We saw in Section 1 that when X and Y are Hilbert spaces, the eigenvectors e_n of the absolute value $S = (T^*T)^{1/2}$ of T in the Schmidt representation (2.1.2) are such that (e_n) is an orthonormal basis of the orthogonal complement $(\ker S)^\perp$ of S, and hence a basis of X if $\ker T = \ker S = \{0\}$. Our main objective in this section is to investigate the properties of the sequence (x_n) of j-eigenvectors of T when X are Banach spaces. The celebrated result of Per Enflo in [45] on the existence of Banach spaces without the approximation property, and consequently without a basis, is a barrier to achieving a result which is comparable to that for Hilbert spaces (see Remark 2.2.33 below), but we shall see that progress is possible if it is assumed that the sequence (S_n) of linear projections S_n is bounded, that is, $\sup_{n \in \mathbb{N}}\{\|S_n\|\} < \infty$. Circumstances in which this condition does or does not hold will also be investigated, as well as further implications of the condition. The assumptions that $T : X \to Y$ is a compact linear map and X, Y are real reflexive spaces which are strictly convex and have strictly convex duals hold throughout the subsection.

Lemma 2.2.22. *Suppose that (S_n) is bounded (equivalently, for all $x \in X$, $(S_n x)$ is bounded). Then for all $x \in X$, $S_n x \rightharpoonup x$ in X/X_∞.*

Proof. Since $S_n x - x \in X_n \subset X_k = {}^0 M_{k-1}$ if $n > k$, we see that for all $y \in \cup_{k \in \mathbb{N}} M_k$,

$$\langle S_n x - x, y \rangle_X \to 0 \quad \text{as} \quad n \to \infty.$$

The boundedness of (S_n) implies that this holds for all $y \in \overline{\cup_{k \in \mathbb{N}} M_k} = (X_\infty)^0$ (see (2.2.48)). Hence, with $\phi_\infty : X \to X/X_\infty$ denoting the canonical map,

$$\langle \phi_\infty(S_n x - x), (\phi_\infty^*)^{-1} y \rangle_{X/X_\infty} \to 0, \quad \text{as} \quad n \to \infty$$

for all $y \in X_\infty^0$, which yields the result since $(\phi^*)^{-1} X_\infty^0 = (X/X_\infty)^*$ by Proposition 2.2.1. $\square$

Corollary 2.2.23. *Suppose that (S_n) is bounded. Then (x_n) is a weak basis, and hence a basis, of X/X_∞. Furthermore, (ξ_n) is a basis of X_∞^0. In particular, if $\ker(T) = \{0\}$, (x_n) and (ξ_n) are bases of X, X^* respectively. Hence, if $\ker(T) = \{0\}$,*

$$x = \sum_{j=1}^{\infty} \xi_j(x) x_j, \quad Tx = \sum_{j=1}^{\infty} \lambda_j \xi_j(x) y_j, \quad y_j = Tx_j/\lambda_j,$$

for all $x \in X$.

Proof. Let $x \in X/X_\infty$. Then by Lemma 2.2.22,

$$x = \sum_{j=1}^{\infty} \xi_j(x) x_j$$

in the sense of weak convergence in X/X_∞. Suppose that there is another (weak) representation $x = \sum_{j=1}^\infty \eta_j(x)x_j$, say. Then since $(x_l, x_k)_X = 0$ if $l < k$, and $(x_1, y)_X = 0$ for all $y \in X_\infty$,

$$0 = \left(x_1, \sum_{j=1}^\infty (\xi_j(x) - \eta_j(x))x_j\right)_X = \xi_1(x) - \eta_1(x).$$

Equality of the other coefficients is proved in the same way, and so (x_n) is a weak basis of X/X_∞. That it is also a basis of X/X_∞ follows from Theorem 1.2.15.

Since X, and hence X/X_∞, are reflexive, it follows from Theorem 1.2.13 that (x_n) is shrinking. The sequences (x_n), (ξ_n) constitute a biorthogonal system in view of Lemma 2.2.9 and consequently (ξ_n) is a basis of $(X/X_\infty)^* \simeq X_\infty^0$, by Proposition 1.2.11. $\qquad\square$

Theorem 2.2.24. *If X is uniformly convex and X^* is strictly convex, then $X = X_\infty \uplus \overline{\bigcup_{n\in\mathbb{N}} J_X^{-1} M_{n-1}}$.*

Proof. Since for any $x \in X$ and $n \in \mathbb{N}$, we have from (2.2.41) that $(I - P_n)x \in J_X^{-1} M_{n-1}$, it follows from the representation $x = P_n x + (I - P_n)x$ and Lemma 2.2.12 that any $x \in X$ can be written as

$$x = P_\infty x + z, \quad z \in \overline{\bigcup_{n\in\mathbb{N}} J_X^{-1} M_{n-1}}.$$

If $x \in X_\infty \cap \left[\overline{\bigcup_{n\in\mathbb{N}} J_X^{-1} M_{n-1}}\right]$, then, since the subspaces M_{n-1} are increasing with n, we have that $x = \lim_{n\to\infty} z_n$, where $z_n \in J_X^{-1} M_{n-1}$. Then $J_X z_n$ converges weakly to $J_X x$ (see Proposition 1.1.26) and so

$$\langle x, J_X x\rangle_X = \lim_{n\to\infty} \langle x, J_X z_n\rangle_X = 0,$$

since $x \in X_n =^0 M_{n-1}$ for every $n \in \mathbb{N}$ and $J_X z_n \in M_{n-1}$.

Finally, let $y \in X_\infty$ and $x \in \bigcup_{n\in\mathbb{N}} J_X^{-1} M_{n-1}$, with $x = \lim_{n\to\infty} z_n$ as before. Then we again have

$$(x, y)_X := \frac{\|x\|_X}{\mu(\|x\|_X)}\langle y, J_X x\rangle_X = \frac{\|x\|_X}{\mu(\|x\|_X)} \lim_{n\to\infty} \langle y, J_X z_n\rangle_X = 0.$$

The theorem is therefore proved. $\qquad\square$

The boundedness of (S_n) is not required for the following result. We recall from Section 1.2 in Chapter 1 that a sequence (x_n) is called a basic sequence if it is a basis of the closed linear span of the x_n.

Proposition 2.2.25. *Suppose that $\ker(T) = \{0\}$. Then the sequence $(x_n)_{n\in\mathbb{N}}$ converges weakly to 0, and there is a subsequence of $(x_n)_{n\in\mathbb{N}}$ that is a basic sequence.*

Proof. Since $\|x_n\|_X = 1$ for all $n \in \mathbb{N}$, there is a subsequence of $(x_n)_{n \in \mathbb{N}}$, still denoted by $(x_n)_{n \in \mathbb{N}}$ for convenience, that converges weakly, to x, say. As T is compact, $Tx_n \to Tx$. But $\|Tx_n\|_Y = \lambda_n \to 0$; thus $Tx = 0$, so that $x = 0$. Now suppose that the whole sequence $(x_n)_{n \in \mathbb{N}}$ does not converge weakly to 0. Then there is a subsequence that converges weakly to some $x \neq 0$, so that by the first argument we have a contradiction. The existence of a subsequence that is a basic sequence is now a consequence of the Bessaga–Pelczyński selection principle (Theorem 1.2.8). □

Proposition 2.2.26. *Let*

$$E_1 = \left\{ x \in X : \lim_{n \to \infty} S_n x = x \right\}, E_2 = \overline{\cup Z_{n-1}}.$$

Then E_1 is a dense linear subspace of E_2; it is closed (and hence $E_1 = E_2$) if and only if $(S_n \upharpoonright_{E_2})$ is bounded.

Proof. Since $S_n(X) = Z_{n-1}$, E_1 is a linear subspace of E_2. Let $x \in E_2$ and $\varepsilon > 0$. Since the Z_{n-1} are increasing, there exist $N \in \mathbb{N}$ and $z_N \in Z_N$ such that for all $n > N$, $z_N = S_n z_N \in Z_{n-1}$ and $\|x - z_N\|_X < \varepsilon$. Hence $z_N \in E_1$: the density of E_1 follows.

If $(S_n \upharpoonright_{E_2})$ is bounded, then for all $x \in E_2$ and all $n > N$, with the same notation as above we have for some constant C,

$$\|x - S_n x\|_X = \|x - z_N - S_n(x - z_N)\|_X \leq C\varepsilon.$$

Hence $x \in E_1$ and so $E_1 = E_2$. Conversely, if $E_1 = E_2$, then $S_n x \to x$ for all $x \in E_2$, so that $(S_n x)$ is bounded for all $x \in E_2$. The uniform boundedness principle implies that (S_n) is bounded. □

Remark 2.2.27. Since (x_n) is a basis of $Z := \cup Z_{n-1}$, it follows that if

$$\sup_n \|S_n \upharpoonright_Z\| < \infty,$$

then (x_n) is also a basis of $\overline{Z}$, by Lemma 1.2.2, and so

$$\sup_n \|S_n \upharpoonright_{E_2}\| < \infty.$$

Thus in Proposition 2.2.26 the condition $\sup_n \|S_n \upharpoonright_{E_2}\| < \infty$ can be replaced by $\sup_n \|S_n \upharpoonright_Z\| < \infty$.

Remark 2.2.28. Clearly $Z_{n-1} \subset \mathrm{sp}\, J_X^{-1} M_{n-1}$. Put $K_{n-1} = J_X^{-1} M_{n-1}$ and regard K_{n-1} as a subset of F_{n-1}, the closed linear span of K_{n-1} endowed with the norm inherited from X. We shall prove in the following lemma that the subspace F_{n-1} is finite dimensional. If its dimension is $n - 1$ then of course $F_{n-1} = Z_{n-1}$ and in this case, in view of (2.2.41), $(I - P_n)$ has range in Z_{n-1}. Since P_n has range X_n it follows that $(I - P_n) = S_n$, and so $(\|S_n\|)$ is bounded; this is so if X is a

Hilbert space (see Lemma 2.2.10), but is also the case in Example 2.2.35 for l_p below. Hence from Theorem 2.2.14, since $S_n^2 = S_n$,

$$x = \sum_{j=1}^{\infty} \xi_j(x)x_j + P_\infty x.$$

Lemma 2.2.29. *Suppose X is uniformly convex. For each $n \in \mathbb{N}$, the closed linear span F_{n-1} of $J_X^{-1} M_{n-1}$ is of finite dimension. Hence it has the direct sum decomposition*

$$F_{n-1} = Z_{n-1} \bigoplus G_{n-1}$$

for some finite-dimensional subspace G_{n-1} and

$$X = X_\infty \biguplus \overline{\bigcup_{n \in \mathbb{N}} \left(Z_{n-1} \bigoplus G_{n-1} \right)}.$$

Proof. Let (y_j) be a bounded sequence in $K_{n-1} = J_X^{-1} M_{n-1}$. Then $(J_X y_j)$ is a bounded sequence in the finite-dimensional space M_{n-1}, and so some subsequence of it, again denoted by $(J_X y_j)$, converges, to $z \in M_{n-1}$, say. Hence $y_j \to J_X^{-1} z$ by Proposition 1.1.26. Thus K_{n-1} is precompact. In fact, as the inverse image of a closed set under a continuous map, it is closed and hence compact. Therefore the closed balanced hull $\hat{K}_{n-1}$ of K_{n-1} is compact (see [11], p. 80). As a subset of F_{n-1}, the set $\hat{K}_{n-1}$ is convex and contains 0. We now claim that

$$F_{n-1} = \bigcup_{j=1}^{\infty} j\hat{K}_{n-1}.$$

To check this, first suppose that $x \in \operatorname{sp} K_{n-1}$. Then $x = \sum_{k=1}^{n-1} \lambda_k z_k$ for some scalars λ_k and some points $z_k \in K_{n-1}$. Thus, with $\lambda = \sum_{k=1}^{n-1} |\lambda_k|$, we have that $x = \lambda \sum_{k=1}^{n-1} (\lambda_k/\lambda) z_k \in \lambda \hat{K}_{n-1}$, and consequently $\operatorname{sp} K_{n-1} \subset \bigcup_{j=1}^{\infty} j\hat{K}_{n-1}$. Now let $x \in F_{n-1}$. There exists a sequence (w_i) in $\operatorname{sp} K_{n-1}$ that converges to x and is therefore bounded. Hence for some $N, w_i \in N\hat{K}_{n-1}$ for all i, and as $\hat{K}_{n-1}$ is closed, we must have $x \in N\hat{K}_{n-1}$.

The Banach space F_{n-1} is thus expressible as a countable union of closed sets, and so by the Baire category theorem, $\hat{K}_{n-1}$ has non-empty interior. By translating this interior if necessary, it follows that there is an open neighbourhood of 0 in F_{n-1} with compact closure. Hence $\dim F_{n-1} < \infty$; see [83], Theorem 3.12-F. □

Remark 2.2.30. We have shown in Corollary 2.2.23 that if (S_n) is bounded and T has trivial kernel, then (x_j) is a basis of X. The assumption of boundedness on (S_n) can not be omitted, for if X is a space without the **approximation property**, such as that of Theorem 1.2.28, then there are a space Y and a map T with trivial kernel such that (S_n) is not bounded. More generally, if X does not have a basis, then there can be no map T with trivial kernel for which (S_n) is bounded.

It is natural to ask: Is (S_n) bounded whenever X has a basis? The following theorem casts some light on this question.

Theorem 2.2.31. *Let J be a non-empty subset of $\mathbb{N}$ and suppose that $U_J :=$ $\{u_j : j \in J\}$ is a linearly independent set in X. Then there is a set $\{w_j : j \in J\}$ that spans the same subspace of X as U_J and has the semi-orthogonality property $(w_l, w_k)_X = \delta_{l,k}$ $(l \le k)$. If $(u_j) = \{u_j : j \in \mathbb{N}\}$ is a basis of X, then so is $(w_j) = \{w_j : j \in \mathbb{N}\}$, and for $x \in X$ we have that*

$$x = \sum_{j=1}^{\infty} \zeta_j(x) w_j, \qquad (2.2.64)$$

where

$$\zeta_j(x) = \left\langle x - \sum_{i=1}^{j-1} \zeta_i(x) w_i, J_X w_j \right\rangle_X \quad \text{for } j \ge 2, \quad \text{and } \zeta_1(x) = \langle x, J_X w_1 \rangle_X.$$

$$(2.2.65)$$

Proof. A Gram–Schmidt type procedure applies. Define

$$v_1 = u_1, w_1 = v_1 / \|v_1\|,$$

$$v_{n+1} = u_{n+1} - \sum_{k=1}^{n} (w_k, u_{n+1})_X w_k, \quad w_{n+1} = v_{n+1} / \|v_{n+1}\| \text{ for } n \in \mathbb{N}.$$

Obviously $(w_n, w_n)_X = 1$ for all n. Moreover,

$$(w_1, w_2)_X = \|v_2\|_X^{-1} \{(w_1, u_2)_X - (w_1, u_2)_X (w_1, w_1)_X\} = 0.$$

Suppose that for some $n \ge 2$, $(w_r, w_s)_X = \delta_{rs}$ if $r \le s \le n$. Then

$$\|v_{n+1}\|_X (w_1, w_{n+1})_X = (w_1, u_{n+1})_X - \sum_{k=1}^{n} (w_k, u_{n+1})_X (w_1, w_k)_X$$

$$= -\sum_{k=2}^{n} (w_k, u_{n+1})_X (w_1, w_k)_X = 0$$

by the inductive hypothesis. Now suppose that $(w_r, w_{n+1})_X = 0$ for all r with $1 \le r \le m < n$. Then

$$\|v_{n+1}\|_X (w_{m+1}, w_{n+1})_X = (w_{m+1}, u_{n+1})_X - \sum_{k=1}^{n} (w_k, u_{n+1})_X (w_{m+1}, w_k)_X$$

$$= -\sum_{k=1, k \ne m+1}^{n} (w_k, u_{n+1})_X (w_{m+1}, w_k)_X.$$

This last sum is zero, for if $k \leq m$, then $(w_k, u_{n+1})_X = 0$ by our assumption, while if $m + 1 < k \leq n$, then $(w_{m+1}, w_k)_X = 0$ by the inductive hypothesis. It follows that $(w_r, w_{n+1})_X = 0$ for all r with $1 \leq r \leq n$, which completes the inductive argument and establishes the claimed semi-orthogonality property.

Subspaces W_n and linear maps R_n can now be defined in terms of the sequence (w_i) just as the X_n and S_n were defined by the sequence (x_i) in Sections 2.2.1 and 2.2.3: the operators R_n are defined by the condition that $x - R_n x \in W_n$ for all $x \in X$ and are uniquely given by

$$R_n x = \sum_{j=1}^{n-1} \zeta_j(x) w_j. \tag{2.2.66}$$

where the $\zeta_j(x)$ are given by (2.2.65); cf. (2.2.18) and (2.2.30).

Let

$$E_n = \mathrm{sp}\{u_1, \ldots, u_{n-1}\}, \quad F_n = \overline{\mathrm{sp}\{u_n, u_{n+1}, \ldots\}}.$$

Then since the u_i form a basis, any $x \in X$ has a unique representation $x = \sum_{i=1}^{\infty} a_i u_i$ say, and hence X has the direct sum decomposition

$$X = E_n \oplus F_n,$$

given by

$$x = \sum_{i=1}^{n-1} a_i u_i + \sum_{i=n}^{\infty} a_i u_i.$$

But since E_n is also spanned by $\{w_j : j = 1, \ldots, n-1\}$, R_n is the (linear) projection of X onto E_n and so we must have that

$$R_n x = \sum_{i=1}^{n-1} a_i u_i \to x$$

as $n \to \infty$. Consequently the semi-orthonormal sequence (w_i) is a basis of X and (2.2.64) holds. $\qquad\square$

It is important to realise that while we have proved in Theorem 2.2.31 that if the space X has a basis then it has a semi-orthonormal basis (w_j) with respect to which (2.2.64) holds, it is not established that this semi-orthonormal basis can be selected to be the sequence (x_n) of j-eigenvectors of the compact linear operator T. Thus whether or not (S_n) is bounded if X has a basis is still an open problem.

The following proposition has interesting implications with respect to the pathological possibilities noted in Remark 2.2.30. We suppose that $X_\infty = \{0\}$ purely for simplicity.

Proposition 2.2.32. *Let S be the space of scalar sequences $\alpha := (\alpha_i)_{i \in \mathbb{N}}$ such that*

$$\|\alpha\|_S := \sup_n \left\| \phi_n \left(\sum_{i=1}^{n-1} \alpha_i x_i \right) \right\|_{X/X_n} < \infty, \tag{2.2.67}$$

and endow it with the norm $\| \cdot \|_S$. Suppose that $X_\infty = \{0\}$. Then the map

$$\Phi : x \mapsto (\xi_i(x)) .$$

is an isometric isomorphism of X onto S.

If the sequence (S_n) is bounded, with $\|S_n\| \leq K$, say, then, for all $m \in \mathbb{N}$ and $x \in X$,

$$\|S_m x\| \leq K \sup_n \left\| \phi_n \left(\sum_{i=1}^{n-1} \xi_i x_i \right) \right\|_{X/X_n} . \tag{2.2.68}$$

Proof. The map Φ is obviously linear. Also Φ maps X into S and is an isometry, for

$$\|(\xi_i(x))\|_S = \sup_n (\|\phi_n(S_n x)\|_{X/X_n}) = \sup_n (\|\phi_n(x)\|_{X/X_n}) = \|x\|_X,$$

by (2.2.49). It remains only to prove that Φ is surjective. To do this, let $\alpha = (\alpha_i) \in S$ and for each $n \in \mathbb{N}$ with $n > 1$, let z_n be the element in $\phi_n \left(\sum_{i=1}^{n-1} \alpha_i x_i \right)$ with minimum X norm. Thus

$$z_n = (I - P_n) \left(\sum_{i=1}^{n-1} \alpha_i x_i \right)$$

and

$$\|z_n\|_X = \left\| \phi_n \left(\sum_{i=1}^{n-1} \alpha_i x_i \right) \right\|_{X/X_n} \leq \|\alpha\|_S .$$

Hence (z_n) is a bounded sequence in X and it contains a subsequence (call it (z_n) still) which is weakly convergent to a limit z in X.

For $m > n$, we have

$$z_m = \sum_{i=1}^{n-1} \alpha_i x_i + \sum_{i=n}^{m-1} \alpha_i x_i - P_m \left(\sum_{i=1}^{m-1} \alpha_i x_i \right) . \tag{2.2.69}$$

Since $\langle x_j, J_X x_i \rangle = 0$ for $i < j$ it follows that $\sum_{i=n}^{m-1} \alpha_i x_i \in X_n$. Hence, as $X_m \subset X_n$, we have

$$\phi_n(z_m) = \phi_n \left(\sum_{i=1}^{n-1} \alpha_i x_i \right), \tag{2.2.70}$$

and so, since $\phi_n(z_m)$ converges weakly to $\phi_n(z)$ as $m \to \infty$,

$$\phi_n(z) = \phi_n \left(\sum_{i=1}^{n-1} \alpha_i x_i \right). \tag{2.2.71}$$

But

$$\phi_n(z) = \phi_n(S_n z) = \phi_n \left(\sum_{i=1}^{n-1} \xi_i(z) x_i \right). \tag{2.2.72}$$

Hence

$$\sum_{i=1}^{n-1} (\alpha_i - \xi_i(z)) x_i \in X_n$$

But, on using (2.2.15), we see that this implies that $\alpha_i = \xi_i(z)$ for $i = 1, \ldots, n-1$ and hence $\alpha = \Phi z$. Thus Φ is surjective.

If $\|S_n\| \le K$, then

$$\|S_m x\|_X \le K \|x\|_X = K \|\Phi x\|_{\mathcal{S}}$$

and (2.2.68) follows. □

To summarise the position, we have

Remark 2.2.33. The following statements are equivalent:

1. (x_n) is a basis of X;
2. $\sup_{n \in \mathbb{N}} \|S_n\| < \infty$;
3. the canonical elements $\delta_n := (\delta_{j,n})_{j \in \mathbb{N}}$ form a basis of $\mathcal{S}$.

Suppose that (x_n) is a basis of X, so that $K := \sup_{n \in \mathbb{N}} \|S_n\| < \infty$. Then in Proposition 2.2.32, for any $x \in X$, since $X_n \supset X_\infty = \{0\}$,

$$\|(\xi_i(x))\|_{\mathcal{S}} = \sup_{n \in \mathbb{N}} \|\phi_n(S_n x)\|_{X/X_n}$$
$$= \sup_{n \in \mathbb{N}} \|\phi_n x\|_{X/X_n} = \|x\|_X$$

and

$$\|x\|_X = \lim_{n \to \infty} \|S_n x\|_X \le \sup_{n \in \mathbb{N}} \|S_n x\|_X.$$

Also,

$$\sup_{n \in \mathbb{N}} \|S_n x\|_X \le K \|x\|_X = K \|(\xi(x))\|_{\mathcal{S}}.$$

Hence, setting $\|(\xi_n(x))\|_{\mathcal{S}'} := \sup_{n \in \mathbb{N}} \|S_n x\|_X$, we have

$$(1/K)\|(\xi_n(x))\|_{\mathcal{S}'} \le \|(\xi_n(x))\|_{\mathcal{S}} \le \|(\xi_n(x))\|_{\mathcal{S}'}.$$

It therefore follows from Proposition 2.2.32 that if (x_n) is a basis of X, then X and $\mathcal{S}$ are linearly homeomorphic to the space $\mathcal{S}'$ of sequences (a_n) with norm

$$\|(a_n)\|_{\mathcal{S}'} := \sup_{n \in \mathbb{N}} \left\| \sum_{j=1}^{n} a_n x_n \right\|_X .$$

This recovers a special case of Proposition 6.9 in [46]: in that proposition, (x_n) is allowed to be any shrinking basis of an arbitrary Banach space X, and $\mathcal{S}'$ is shown to be isomorphic to X^{**}, the isomorphism being also an isometry if (x_n) is a monotone basis.

Remark 2.2.34. Let X be the subspace of $l_p, 2 < p < \infty$, without the approximation property, constructed in Theorem 1.2.28. It does not contain all the canonical basis elements, but we see from Proposition 2.2.32 and Lemma 2.2.9 that it is isometrically isomorphic to the sequence space $\mathcal{S}$ which does contain all the sequences δ_n. Note that the canonical elements δ_n in $\mathcal{S}$ inherit the semi-orthogonality property from (x_n). This follows on applying Proposition 2.2.32 to Definition 1.1.30 and Proposition 1.1.31. Of course, the semi-orthogonality in $\mathcal{S}$ is understood in the sense of the inner product $(\cdot, \cdot)_{\mathcal{S}}$ defined as in Proposition 1.1.31 with supporting functional $\widetilde{J}_{\mathcal{S}}$.

Example 2.2.35. Let $X = Y = l_p (1 < p < \infty)$ and $T : l_p \to l_p$ the diagonal map $T(a^{(j)}) = (\sigma^{(j)} a^{(j)})$, where the $\sigma^{(j)}$ are non-negative, decreasing and converge to 0 as $j \to \infty$. Then T is compact and $\|T\| = \sigma^{(1)}$. The duality map J_p on l_p corresponding to the choice of gauge function $\mu(t) = t^{p-1}$ is given by

$$J_p z = \left(|z^{(j)}|^{p-2} z^{(j)} \right). \tag{2.2.73}$$

Furthermore we have the following for $n \in \mathbb{N}$:

$$x_n = \{\delta_{jn}, j \in \mathbb{N}\}, \quad J_p x_n = x_n,$$
$$X_n = \{z = (z^{(j)}) \in l_p : z^{(j)} = 0 \text{ for } j < n\} \tag{2.2.74}$$
$$\xi_n(z) = z^{(n)}, \quad \lambda_n = \sigma^{(n)}.$$

Therefore $S_n z = \sum_{j=1}^{n-1} z^{(j)} x_j$ and $\|S_n z\|_{l_p} \leq \|z\|_{l_p}$. Moreover, $I - P_n = S_n$ and $Z_{n-1} = J_p^{-1} M_{n-1}$.

We reiterate that if X is the subspace of l_p without the approximation property (and hence without a basis) constructed in Theorem 1.2.28, it does not contain the canonical basis elements. Taking X to be this subspace and T_X to be the restriction of T to X, the j-eigenvectors of T_X do not form a basis of X, and hence the corresponding sequence (S_n) is not bounded.

Besov spaces of the form $B_{p,p}^s(\Omega)$ are isomorphic to certain weighted sequence spaces $b_{p,p}^s$, and in this way consideration of the natural embedding of $B_{p,p}^{s_1}(\Omega)$ in

$B_{p,p}^{s_2}(\Omega)$ ($s_1 > s_2$) reduces to consideration of a diagonal map from $b_{p,p}^{s_1}$ to $b_{p,p}^{s_2}$, the only effect of the diagonal map being to change the weights. If we chose to identify the Besov spaces with these sequence spaces, then we would have boundedness of the corresponding S_n.

Remark 2.2.36. To determine whether or not $(S_n)_{n \in \mathbb{N}}$ is bounded is difficult in general. It is therefore natural to ask if a topology can be imposed on X with respect to which $(S_n)_{n \in \mathbb{N}}$ is bounded. The **projective limit** fulfills this role, being the coarsest topology on X compatible with the algebraic structure of X under which the maps S_n are continuous; it is a locally convex topology; see [78].

Suppose that $X_\infty = \{0\}$. Then

$$S_n : X \to Z_{n-1} = \mathrm{sp}\{x_1, x_2, \ldots, x_{n-1}\} \simeq X/X_n$$

implies that

$$\bigcap_{n \in \mathbb{N}} S_n^{-1}(\{0\}) = \bigcap_{n \in \mathbb{N}} X_n = \{0\}.$$

If $V_n := \{v_n\}$ is a base of absolutely convex neighbourhoods in Z_{n-1}, then finite intersections of $S_n^{-1} v_n$ ($v_n \in V_n$) form a base of absolutely convex neighbourhoods of X in the projective limit topology.

We have $v_n = B_\varepsilon(0) \cap Z_{n-1}$ for some $\varepsilon > 0$, where $B_\varepsilon(0)$ is the ball centre 0 and radius ε defined with respect to the norm induced by X (which is equivalent to any other norm since Z_{n-1} is of finite dimension). Since

$$S_n^{-1}\left(B_\varepsilon(0) \cap Z_{n-1}\right) = \{B_\varepsilon(0) \cap Z_{n-1} + X_n\}$$

and

$$\{B_\varepsilon(0) \cap Z_{n-1} + X_n\} \bigcap \{B_\delta(0) \cap Z_{k-1} + X_k\}$$
$$\supset \left\{B_{\min\{\varepsilon,\delta\}}(0) \cap Z_{\min\{n,k\}-1} + X_{\max\{n,k\}}\right\},$$

a base of neighbourhoods of X is given by sets of the form

$$\{B_\varepsilon(0) \cap Z_{m-1} + X_k : \varepsilon > 0, k \geq m\}.$$

Given $\varepsilon > 0, k \in \mathbb{N}$, then, for $n \geq k$,

$$S_n(x) - x \in X_n \subset B_\varepsilon(0) \cap Z_{m-1} + X_k.$$

Therefore $S_n x \to x$ in the projective limit topology on X.

2.2.6 A Schmidt-type expansion for T

We continue to make our basic assumptions that $T : X \to Y$ is a compact linear map between the real, reflexive Banach spaces X, Y which are strictly convex and have strictly convex duals. In Corollary 2.2.23, we proved that if (S_n) is bounded and T has trivial kernel, then T has the representation

$$Tx = \sum_{j=1}^{\infty} \lambda_j \xi_j(x) y_j, \quad y_j = Tx_j/\lambda_j.$$

Our goal in this section is to present a result established in [41], that this representation is valid without the assumption that (S_n) is bounded, as long as it is assumed that the j-eigenvalues λ_n decay sufficiently fast at infinity.

We shall make use of the linear projections $R_k : Y \to \mathrm{sp}\{y_1, y_2, \ldots, y_{k-1}\}$ determined by the conditions $y - R_k y \in Y_k$ for all $y \in Y$ and given by (2.2.53) and (2.2.54); we recall from (2.2.8) and (2.2.9) that $Y_k :=^0 N_{k-1}$, $N_{k-1} := \mathrm{sp}\{J_Y Tx_1, \ldots, J_Y Tx_{k-1}\}$.

Lemma 2.2.37. *Let $n \in \mathbb{N}$ and let K_n^X, K_n^Y be respectively the convex hulls of $x_1, x_2, \ldots, x_n$ and $y_1, y_2, \ldots, y_n$. Then*

$$(2^n - 1)^{-1} \leq \inf\{\|x\| : x \in K_n^X\} \leq 1 \tag{2.2.75}$$

and

$$(2^n - 1)^{-1} \leq \inf\{\|y\| : y \in K_n^Y\} \leq 1. \tag{2.2.76}$$

Proof. Any $x \in K_n^X$ is of the form $x = \sum_{i=1}^{n} \alpha_i x_i$, for some non-negative α_i satisfying $\sum_{i=1}^{n} \alpha_i = 1$. Since $\sum_{i=2}^{n} \alpha_i x_i \in X_2$ and $\alpha_1 x_1 \perp^j X_2$ by (2.2.32), we have that

$$\alpha_1 = \|\alpha_1 x_1\|_X \leq \left\|\alpha_1 x_1 + \sum_{i=2}^{n} \alpha_i x_i\right\|_X = \|x\|_X.$$

As $\sum_{i=3}^{n} \alpha_i x_i \in X_3$ and $\alpha_2 x_2 \perp^j X_3$,

$$\alpha_2 - \alpha_1 = \|\alpha_2 x_2\|_X - \|\alpha_1 x_1\|_X \leq \left\|\alpha_2 x_2 + \sum_{i=3}^{n} \alpha_i x_i\right\|_X - \|\alpha_1 x_1\|_X$$

$$\leq \left\|\sum_{i=1}^{n} \alpha_i x_i\right\|_X = \|x\|_X.$$

Similarly it follows that for all $l \in \mathbb{N}$,

$$\alpha_{l+1} - \alpha_l \leq \left\|\sum_{i=l}^{n} \alpha_i x_i\right\|_X.$$

Therefore

$$\alpha_3 - \alpha_2 - \alpha_1 \leq \left\|\sum_{i=2}^{n} \alpha_i x_i\right\|_X - \|\alpha_1 x_1\|_X \leq \left\|\sum_{i=1}^{n} \alpha_i x_i\right\|_X = \|x\|_X,$$

and

$$\alpha_4 - \alpha_3 - \alpha_2 - \alpha_1 \leq \left\| \sum_{i=3}^{n} \alpha_i x_i \right\|_X - \|\alpha_2 x_2\|_X - \|\alpha_1 x_1\|_X$$

$$\leq \left\| \sum_{i=2}^{n} \alpha_i x_i \right\|_X - \|\alpha_1 x_1\|_X \leq \left\| \sum_{i=1}^{n} \alpha_i x_i \right\|_X = \|x\|_X.$$

Continuing in this way we obtain, for each $l \in \mathbb{N}$ with $l \leq n$,

$$\alpha_l - \sum_{i=1}^{l-1} \alpha_i \leq \|x\|_X.$$

Thus

$$\inf \ \max\{\alpha_1, \alpha_2 - \alpha_1, \alpha_3 - \alpha_2 - \alpha_1, \ldots, \alpha_n - \alpha_{n-1} - \cdots - \alpha_1\} \leq \inf_{x \in K_n} \|x\|_X, \quad (2.2.77)$$

where the infimum on the left-hand side is taken over all $\alpha_i \geq 0$ such that $\sum_{i=1}^{n} \alpha_i = 1$. When $n = 2$, we have to find $\inf_{0 \leq t \leq 1}(t, 1 - 2t)$. Since

$$\max(t, 1 - 2t) = \begin{cases} 1 - 2t, & 0 \leq t \leq 1/3, \\ t, & 1/3 \leq t \leq 1, \end{cases}$$

we see that the infimum is attained when the two entries are equal, that is, when $t = 1/3$, as otherwise one of the two entries would be greater than this value. For a general value of n, the left-hand side of (2.2.77) is of the form

$$\inf \max\{\alpha_1, \alpha_2 - \alpha_1, \ldots, 1 - 2(\alpha_1 + \cdots + \alpha_{n-1})\}.$$

If the infimum is attained at $\alpha_i - \alpha_{i-1} - \cdots - \alpha_1$, we can decrease α_i by a sufficiently small $\varepsilon > 0$ while increasing each α_j with $j \in \{i+1, \ldots, n\}$ by $\varepsilon/(n-i)$: this leads to a decrease of $\alpha_i - \alpha_{i-1} - \cdots - \alpha_1$ by ε and an increase of every term $\alpha_j - \alpha_{j-1} - \cdots - \alpha_1$ with $j > i$. Considerations like this show that, as before, the infimum is attained when the entries coincide: $\alpha_1 = \alpha_2 - \alpha_1$, so that $\alpha_2 = 2\alpha_1$; $\alpha_3 = 2\alpha_2 = 2^2\alpha_1$ and so on up to $\alpha_{n-1} = 2^{n-2}\alpha_1$. Moreover,

$$\alpha_1 = 1 - 2\alpha_1(1 + 2 + \cdots + 2^{n-2})$$

which gives $\alpha_1 = (2^n - 1)^{-1}$ and hence the lower bound in (2.2.75). The upper bound is obvious and so the proof of (2.2.75) is complete. The proof of (2.2.76) is similar. $\qquad\square$

Lemma 2.2.38. *If $\lambda_n \leq 2^{-n+1}$ for all $n \in \mathbb{N}$, then*

$$\|y - R_n y\|_Y \leq 1$$

for all $y \in T(B_X)$.

Proof. Let $y = Tx$, $x \in B_X$. By Lemma 2.2.21 and (2.2.57),

$$y = R_2 y + Q_2 y = R_2 T x + Q_2 y = T S_2 x + Q_2 y,$$

and so, by Lemma 2.2.10,

$$Q_2 y = T(x - S_2 x) = T P_2 x.$$

Since $\|P_2 x\|_X \le \|P_2 x - x\|_X + \|x\|_X \le 2\|x\|_X$, it follows that

$$\frac{1}{2} Q_2 T(B_X) \subset Y_2 \cap T(B_X). \tag{2.2.78}$$

If $y \in Y_2 \cap T(B_X)$, then $y = Tx$, $x \in X_2 \cap B_X$ by (2.2.58). Hence

$$\sup\{\|y\|_Y : y \in Y_2 \cap T(B_X)\} = \lambda_2 \le 2^{-1}.$$

Further use of Lemma 2.2.21 and (2.2.57) shows, by use of similar techniques, that

$$\frac{1}{2} Q_3 \left(T(B_X) \cap Y_2 \right) \subset T(B_X) \cap Y_2 \cap Y_3,$$

which together with (2.2.78) gives

$$\frac{1}{2} Q_3 \left(\frac{1}{2} Q_2 T(B_X) \right) \subset T(B_X) \cap Y_3.$$

Hence the condition $\lambda_3 \le 2^{-2}$ implies that

$$\sup\{\|y\|_Y : y \in Q_3 Q_2 T(B_X)\} \le 1.$$

More generally, the same procedure shows that for any $n \in \mathbb{N} \setminus \{1\}$, if $\lambda_n \le 2^{-n+1}$, then

$$\sup\{\|y\|_Y : y \in Q_n Q_{n-1} \cdots Q_2 T(B_X)\} \le 1.$$

Together with Lemma 2.2.21, this finishes the proof.   □

Lemma 2.2.38 implies that the R_n, regarded as maps from the closure in Y of TX to itself, have uniformly bounded norms. From this follows the main result

Theorem 2.2.39. *If T has dense range in Y and $\lambda_n \le 2^{-n+1}$ for all $n \in \mathbb{N}$, then for all $x \in X$,*

$$Tx = \sum_{n=1}^{\infty} \lambda_n \xi_n(x) y_n, \quad y_n = \lambda_n^{-1} T x_n.$$

Proof. The proof is similar to that for Corollary 2.2.23. Let $y \in Y$. Since $R_n y - y \in Y_n \subset Y_k =^0 N_{k-1}$ if $n > k$, it follows that for all $z \in \cup_{k \in \mathbb{N}} N_k$,

$$\langle R_n y - y, z \rangle_Y \to 0 \quad \text{as} \quad n \to \infty.$$

By Lemma 2.2.38, (R_n) is bounded and so this limit continues to hold for all $z \in \bigcup_{k \in \mathbb{N}} N_k$. But

$$\overline{\bigcup_{k \in \mathbb{N}} N_k} = (\cap_{k \in \mathbb{N}} Y_k)^0 = Y^*$$

by (2.2.60). Hence $R_n y \rightharpoonup y$ as $n \to \infty$. It follows as in the proof of Corollary 2.2.23 that the consequent weak representation of y is unique and thus (y_n) is a weak basis, and therefore a basis by Theorem 1.2.15. Note that by (2.2.57),

$$Tx = \lim_{n \to \infty} R_n Tx = \lim_{n \to \infty} TS_n x = \sum_{n=1}^{\infty} \lambda_n \xi_n(x) y_n. \qquad \square$$

Since the λ_n are rather difficult to analyse in general, it is clearly desirable that the last theorem be shown to hinge on the decay of more familiar objects associated with T. This was done in [41] where the inequalities in the following lemma were established between the Gelfand widths $\tilde{c}_n(T)$ and λ_n.

Lemma 2.2.40. *For every $n \in \mathbb{N}$,*

$$(2^n - 1)^{-1} \lambda_n \le \tilde{c}_n(T) \le \lambda_n.$$

Proof. Let $n \in \mathbb{N}$. Then $T(B_X)$ contains the convex hull H_n of $\pm \lambda_n y_1, \ldots, \pm \lambda_n y_n$. Let $\varepsilon > 0$. We claim that, with $V_n = \text{sp}\{y_1, \ldots, y_n\}$,

$$T(B_X) \cap V_n \not\supseteq \{y \in Y : \|y\|_Y \le \tilde{c}_n(T) + \varepsilon\} \cap V_n.$$

For if K_n is a linear subspace of Y with codimension $n-1$, then $\dim(K_n \cap V_n) \ge 1$, so that if the claim were false, the definition of $\tilde{c}_n(T)$ would be contradicted. By Lemma 2.2.37 applied to the y_i,

$$H_n \supset \{y \in Y : \|y\|_Y \le \lambda_n(2^n - 1)^{-1}\} \cap V_n,$$

so that

$$(T(B_X) \cap V_n) \supset \{y \in Y : \|y\|_Y \le \lambda_n(2^n - 1)^{-1}\} \cap V_n.$$

It follows that $\tilde{c}_n(T) + \varepsilon > \lambda_n(2^n - 1)^{-1}$, whence the result. $\qquad \square$

In Theorem 1.4.6 it was proved that the Gelfand numbers $c_n(T)$ and Gelfand widths $\tilde{c}_n(T)$ are equal if X and Y are uniformly convex and uniformly smooth, and the operator T has trivial kernel and dense range. Theorem 2.2.39 can now be recast in terms of the decay of the Gelfand numbers, which have been extensively studied.

Theorem 2.2.41. *Let both X and Y be uniformly convex and uniformly smooth, and let T have trivial kernel and dense range. If $c_n(T) \le 2^{-n+1}(2^n - 1)^{-1}$ for all $n \in \mathbb{N}$, then (y_n) is a basis of Y. In particular, for all $x \in X$,*

$$Tx = \sum_{n=1}^{\infty} \lambda_n \xi_n(x) y_n.$$

Proof. Let $y \in Y$. Since $R_n y - y \in Y_n \subset Y_k =^0 N_{k-1}$ if $n > k$, it follows that for all $z \in \cup_{k \in \mathbb{N}} N_k$,

$$\langle R_n y - y, z \rangle_Y \to 0 \quad \text{as} \quad n \to \infty. \tag{2.2.79}$$

Since $c_n(T) = \tilde{c}_n(T)$, the assumption on $c_n(T)$ implies that $\lambda_n \leq 2^{-n+1}$, by Lemma 2.2.40. Hence $\|y - R_n y\|_Y \leq 1$ by Lemma 2.2.38, for all $y \in T(B_X)$ and all $n \in \mathbb{N}$ and so (R_n) is bounded. Consequently (2.2.79) holds for all $z \in \cup_{k \in \mathbb{N}} N_k = Y^*$ and hence $R_n y \rightharpoonup y$. In other words, from (2.2.53),

$$y = \sum_{j=1}^{\infty} \zeta_j(y) y_j$$

in the sense of weak convergence in Y. The uniqueness of this weak representation follows from the semi-orthogonality property $(y_l, y_k)_Y = 0$ if $l < k$, and so $(y_n)_{n \in \mathbb{N}}$ is a weak basis of Y. By Banach's weak basis theorem, it is a basis of Y. Since $\zeta_j(Tx) = \lambda_j \xi(x)$ for all $x \in X$, by (2.2.57), the proof is complete. $\square$

Remark 2.2.42. As a consequence of Proposition 2.5.3 below it will follow that under the assumption of rapid decay that we made about the j-eigenvalues in Theorem 2.2.39, the operator T is nuclear, a notion that will be defined and discussed in Section 2.5 below; such operators are compact and have a series representation. Our representation results in this connection thus apply only to a subset of the class of nuclear maps. Since

$$a_n(T) \leq 2\sqrt{n} c_n(T)$$

(see (1.4.1)), it also follows from the decay assumption in Theorem 2.2.39 that $(a_n(T))_{n \in \mathbb{N}} \in l_1$ which, by a result of Pietsch (see [74], 6.3.3.3, p. 346), again implies the nuclearity of T. The main point of Theorem 2.2.41 is to give a particular decomposition in which the coefficients are recursively calculable by procedures analogous to those in the Hilbert space case.

2.3   Applications

Let λ_1 be the first j-eigenvalue with corresponding j-eigenfunction x_1: thus

$$\|x_1\|_X = 1, \lambda_1 = \|Tx_1\|_Y = \|T\|$$

and

$$T^* J_Y T x_1 = \nu_1 J_X x_1, \quad \nu_1 = \lambda_1 \mu(\lambda_1). \tag{2.3.1}$$

On setting $x_1^* := J_X x_1$, (2.3.1) can be written in the form

$$T^* J_Y T J_X^{-1} x_1^* = \nu_1 x_1^*, \tag{2.3.2}$$

so that ν_1 is an eigenvalue of the nonlinear operator $T^* J_Y T J_X^{-1} : X^* \to X^*$, with corresponding eigenvector x_1^*. When X and Y are Hilbert spaces, the natural gauge functions are $\mu_X(t) = \mu_Y(t) = t$, so that $\nu_1 = \|T\|^2$ is an eigenvalue of T^*T and hence $\|T\|$ is a singular value of T. We now examine the natural (Carathéodory) and variational ways of defining eigenvalues and eigenfunctions in examples involving the p-Laplacian, and discuss the role of the j-eigenvalues and j-eigenfunctions.

2.3.1 The p-Laplacian

Let Ω be a bounded open subset of $\mathbb{R}^n$ and let $W_p^1(\Omega), 1 < p < \infty$, be the Sobolev space of all real-valued functions $u \in L_p(\Omega)$ all of whose first-order distributional derivatives $D_j u$ also belong to $L_p(\Omega)$. The norm on $W_p^1(\Omega)$ is defined to be

$$\left(\int_\Omega \left\{ |u|^p + \sum_{j=1}^n |D_j u|^p \right\} d\mathbf{x} \right)^{1/p}.$$

We take X to be $\overset{0}{W}{}_p^1(\Omega)$, the closure in $W_p^1(\Omega)$ of the set $C_0^\infty(\Omega)$ of all infinitely differentiable functions with compact support in Ω, and define the norm on X by

$$\|u\|_X = \left(\int_\Omega \sum_{j=1}^n |D_j u|^p \, d\mathbf{x} \right)^{1/p}. \tag{2.3.3}$$

Because of the Friedrichs inequality

$$\int_\Omega |u|^p \, d\mathbf{x} \le \text{const} \int_\Omega \sum_{j=1}^n |D_j u|^p \, d\mathbf{x}$$

(see [32], Theorem V.3.22), this norm is equivalent to the norm on X inherited from $W_p^1(\Omega)$. Let $Y = L_p(\Omega)$ and $T = \text{id}$ the identity map from X into Y, which is well known to be compact. It is plain that both X and Y are reflexive and strictly convex. Obviously Y^* is strictly convex; that the same holds for X^* follows from the observation that $\|\cdot\|_X$ is Gâteaux-differentiable on $X \backslash \{0\}$. Direct verification shows that

$$\tilde{J}_Y u = \|u\|_p^{-(p-1)} |u|^{p-2} u, \tag{2.3.4}$$

where $\|\cdot\|_p$ is the usual norm on $L_p(\Omega)$. As for $\tilde{J}_X$, we claim that

$$\tilde{J}_X u = -\|u\|_X^{-(p-1)} \Delta_p u \text{ in the sense of distributions,} \tag{2.3.5}$$

where

$$\Delta_p u = \sum_{j=1}^n D_j \left(|D_j u|^{p-2} D_j u \right), \tag{2.3.6}$$

corresponding to a version of the p-Laplacian. To verify this, note that for all $u \in X$,

$$\left\langle u, -\|u\|_X^{-(p-1)} \Delta_p u \right\rangle_X = -\|u\|_X^{-(p-1)} \langle u, \Delta_p u \rangle_X$$

$$= \|u\|_X^{-(p-1)} \int_\Omega \sum_{j=1}^n D_j u. |D_j u|^{p-2} D_j u dx$$

$$= \|u\|_X.$$

With $\mu_X(t) = \mu_Y(t) = t^{p-1}$, the corresponding duality maps J_X, J_Y are given by

$$J_X(u) = -\Delta_p u, \quad J_Y(u) = |u|^{p-2} u \tag{2.3.7}$$

and in (2.3.2), $\nu_1 = \lambda_1^p$ where $\lambda_1 = \|T\|$. The Euler equation $T^* J_Y T u_1 = \nu_1 J_X u_1$, is equivalent to

$$\int_\Omega \phi |u_1|^{p-2} u_1 dx = \lambda_1^p \int_\Omega \sum_{j=1}^n (D_j \phi) |D_j u_1|^{p-2} D_j u_1 dx, \quad (\forall \phi \in \overset{0}{W}_p^1(\Omega)) \tag{2.3.8}$$

so that u_1 is a weak solution of the Dirichlet eigenvalue problem

$$-\Delta_p u_1 = \lambda_1^{-p} |u_1|^{p-2} u_1, \ u_1 = 0 \text{ on } \partial\Omega. \tag{2.3.9}$$

Since id is not of finite rank, our general procedure summarised in Theorem 2.2.5 ensures that for each $k \in \mathbb{N}$, there are a j-eigenvector u_k and a corresponding j-eigenvalue λ_k that satisfy

$$-\Delta_p u_k = \lambda_k^{-p} |u_k|^{p-2} u_k, \ u_k = 0 \text{ on } \partial\Omega. \tag{2.3.10}$$

in the sense that for all $\phi \in X_k$,

$$\int_\Omega \phi |u_k|^{p-2} u_k dx = \lambda_k^p \int_\Omega \sum_{j=1}^n (D_j \phi) |D_j u_k|^{p-2} D_j u_k dx; \tag{2.3.11}$$

we shall say that (2.3.10) is satisfied in the k-**weak sense**. Note that when $k = 1$ all functions in $X_1 = X$ are allowed as test functions, so that u_1 is a weak solution of the Dirichlet problem in the conventional sense. However, when $k > 1$ the only test functions allowed are the elements of X_k, which is a proper subset of X, and so the u_k need not be weak solutions in the classical sense. It would be interesting to know what regularity properties are possessed by these k-weak solutions. The j-eigenvalues have the variational representation

$$\lambda_k = \|u_k\|_Y = \sup_{u \in X_k \setminus \{0\}} \left\{ \frac{\|u\|_Y}{\|u\|_X} \right\}$$

and so

$$\lambda_k^{-p} = \inf_{u \in X_k \setminus \{0\}} \left\{ \frac{\int_\Omega \sum_{j=1}^n |D_j u|^p d\mathbf{x}}{\int_\Omega |u|^p d\mathbf{x}} \right\}. \tag{2.3.12}$$

Information about the growth of the λ_k^{-p} can be obtained without difficulty. For each $k \in \mathbb{N}$, the restriction id_k of the embedding $id : \overset{0}{W_p^1}(\Omega) \to L_p(\Omega)$ to X_k coincides with the restriction to X_k of the embedding $Id : W_p^1(\Omega) \to L_p(\Omega)$. Thus from the definition of the Gelfand and Weyl numbers we have $\lambda_k \ge c_k(id_k) \ge c_k(Id) \ge w_k(Id)$. From [59], Theorem 3.c.5 and Remark 3.c.7 (1), we see that if Ω is minimally smooth (see [59], p. 170; also [32], p. 255), then $w_k(Id) \ge ck^{-1/n}$, where c is a positive constant independent of k. Hence $\lambda_k \ge ck^{-1/n}$ and so the j-eigenvalues λ_k^{-p} are $0(k^{p/n})$. This upper estimate of the growth of the j-eigenvalues is exactly that obtained for the Lusternik–Schnirelmann eigenvalues in [47] and [48] (these corresponding to classical weak solutions), where lower bounds of the same order are also established. (See below for further details of Lusternik–Schnirelmann eigenvalues.)

A similar analysis applies to the problem in which $X = \overset{0}{W_p^1}(\Omega)$ and $Y = L_q(\Omega)$, with $1 < p < n$, $1 < q < np/(n-p)$, when $n > 1$; and $p, q \in (1, \infty)$ when $n = 1$. In these circumstances, the natural embedding $T = id : \overset{0}{W_p^1}(\Omega) \to L_q(\Omega)$ is compact; in the case $n = 1$, $\overset{0}{W_p^1}(\Omega)$ is compactly embedded in $C(\overline{\Omega})$, by Remarks 5.8.4 in [60], and hence in $L_q(\Omega)$, for any $q \in (1, \infty)$. The above procedure establishes the existence of a countable family of j-eigenvectors v_k corresponding to j-eigenvalues μ_k that satisfy the p, q-Laplacian Dirichlet eigenvalue problem

$$-\Delta_p v_k = \mu_k^{-q} |v_k|^{q-2} v_k, \quad v_k = 0 \quad \text{on } \partial\Omega$$

in the k-weak sense as before, with

$$\mu_k = \sup_{u \in X_k \setminus \{0\}} \frac{\|Tu\|_Y}{\|u\|_X}.$$

Two special cases are of particular interest. Firstly, if $p = 2$, $n > 2$, $q \in (1, 2n/n - 2)$, the projections P_n onto X_n are given by $P_n = I - S_n$ in view of Lemma 2.2.10, and hence the sequence (S_n) is bounded. Consequently the j-eigenvectors $v_n, n \in \mathbb{N}$, form an orthonormal basis of $X = \overset{0}{W_2^1}(\Omega)$ and for all $x \in X$,

$$x = \sum_{n \in \mathbb{N}} \xi_n(x) v_n \quad \text{in } X,$$

with $\xi_n(x) = (x, v_n)_X$, the inner product on $\overset{0}{W_2^1}(\Omega)$. The equation $-\Delta v = \mu v^b$ is the Emden–Fowler equation which is of importance in astrophysics; the physical conditions require $v \ge 0$.

Secondly, when $q = 2$ and $p > 2n/n + 2$, Corollary 2.2.20 applies, as id has dense range. Since the linear projections R_k in (2.2.53) are now such that (R_k) is bounded we therefore have that, with $Tv_n = \mu_n y_n$, the sequence $\{y_n : n \in \mathbb{N}\}$ is an orthonormal basis of $L_2(\Omega)$ and for all $x \in X$,

$$Tx = \sum_{n \in \mathbb{N}} \mu_n \xi_n(x) y_n \quad \text{in } Y,$$

and $\mu_n \xi_n(x) = (Tx, y_n)_Y$. As $T = \text{id}$, we have a series representation, converging in $L_2(\Omega)$, of every element of $\overset{0}{W}{}^1_p(\Omega)$. Furthermore, $\mu_n |\xi_n(x)| \leq \|Tx\|_Y$, and since $\|S_n x\|_X \leq \sum_{j=1}^\infty |\xi_j(x)|$, it follows that $\|S_n x\|_X \leq \sum_{j=1}^\infty \mu_j^{-1} \|Tx\|_Y$. Hence

$$\sup_{n \in \mathbb{N}} \|S_n\| \leq \gamma \sum_{j=1}^\infty \mu_j^{-1},$$

where γ is the norm of the embedding $X \hookrightarrow Y$, and so $(\|S_n\|)$ is bounded if $(\mu_n^{-1}) \in l_1$. It is interesting to compare this with the general result in Theorem 2.2.39

2.3.2 A weighted problem for the p-Laplacian

We continue to suppose that Ω is a bounded open subset of $\mathbb{R}^n$, $p \in (1, \infty)$, but now introduce functions a, b which are non-negative a.e. on Ω and satisfy

$$a, a^{-1/(p-1)} \in L_1^{loc}(\Omega), \tag{2.3.13}$$

$$a^{-s} \in L_1(\Omega) \quad \text{for some} \quad s \in \left(\frac{n}{p}, \infty\right) \cap \left[\frac{1}{p-1}, \infty\right), \quad ps < n(s+1), \tag{2.3.14}$$

and for some $q \in \left[p, \frac{nps}{n(s+1)-ps}\right)$,

$$b \in L_{q/(q-p)}(\Omega) \text{ if } q \in \left(p, \frac{nps}{n(s+1)-ps}\right), \quad b \in L_\infty(\Omega) \quad \text{if} \quad q = p. \tag{2.3.15}$$

Let X be the completion of $C_0^\infty(\Omega)$ with respect to the norm

$$\left(\int_\Omega \left(|u|^p + a \sum_{j=1}^n |D_j u|^p\right) d\mathbf{x}\right)^{1/p}.$$

Let $p_s = ps/(s+1) < p$, put $p_s^* = nps/(n-p_s)$ and note that under the conditions on s, $p_s^* > p$. By the classical Sobolev embedding theorem (see [32], Theorem

V.3.6), $\overset{0}{W}{}^1_{p_s}(\Omega)$ is continuously embedded in $L_{p_s^*}(\Omega)$; together with the use of Hölder's inequality this gives for all $u \in C_0^\infty(\Omega)$,

$$\left(\int_\Omega |u(x)|^p \, dx \right)^{1/p} \le c_1 \left(\int_\Omega |u(x)|^{p_s^*} \, dx \right)^{1/p_s^*}$$

$$\le c_2 \left(\int_\Omega \left(|u(x)|^{p_s} + \sum_{j=1}^n |D_j u|^{p_s} \right) dx \right)^{1/p_s}.$$

Now use the classical Friedrichs inequality

$$\int_\Omega |u|^p dx \le c \int_\Omega \left(\sum_{j=1}^n |D_j u|^p \right) dx$$

(see [32], Theorem V.3.22) plus Hölder's inequality to obtain

$$\left(\int_\Omega |u(x)|^p \, dx \right)^{1/p} \le c_3 \left(\int_\Omega \sum_{j=1}^n |D_j u|^{p_s} dx \right)^{1/p_s}$$

$$\le c_3 \left(\int_\Omega a^{-s}(x) dx \right) \left(\int_\Omega a(x) \sum_{j=1}^n |D_j u|^p dx \right)^{1/p}.$$

We thus have the weighted Friedrichs inequality (1.29) in [27], namely,

$$\int_\Omega |u(x)|^p \, dx \le c \int_\Omega a(x) \sum_{j=1}^n |D_j u(x)|^p \, dx.$$

A norm on X equivalent to that given earlier is therefore

$$\|u\|_X := \left(\int_\Omega a \sum_{j=1}^n |D_j u|^p dx \right)^{1/p}; \qquad (2.3.16)$$

we use this norm from now on. Then X is reflexive and X and X^* are strictly convex; these properties can be verified as in the case of the p-Laplacian. We take Y to be the weighted Lebesgue space $L_p(\Omega; b)$ defined by means of the norm

$$\|u\|_Y := \left(\int_\Omega |u|^p b \, dx \right)^{1/p}; \qquad (2.3.17)$$

this is also reflexive, strictly convex and has a strictly convex dual. Since X is continuously embedded in $\overset{0}{W}{}^{1}_{p_s}(\Omega)$ which, by the classical Sobolev embedding theorem (see Theorem V.3.7 in [32]) is compactly embedded in $L_q(\Omega)$, it follows that X is compactly embedded in $L_q(\Omega)$. Also, as

$$\int_\Omega b|u|^p d\mathbf{x} \leq \left(\int_\Omega |u|^q d\mathbf{x} \right)^{p/q} \left(\int b^{q/(q-p)} d\mathbf{x} \right)^{(q-p)/p}$$

for $q > p$ (with the natural interpretation if $q = p$), we see that $L_q(\Omega)$ is continuously embedded in Y. Therefore X is compactly embedded in Y and this embedding is taken to be our operator T. The duality maps J_X and J_Y can easily be seen to be given by

$$J_X u = -\sum_{j=1}^n D_j \left(a|D_j u|^{p-2} D_j u \right)$$

and

$$J_Y u = b|u|^{p-2} u.$$

Our procedure establishes the existence of a countable family of j-eigenvectors and j-eigenvalues of the Dirichlet eigenvalue problem

$$-\sum_{j=1}^n D_j \left(a|D_j u|^{p-2} D_j u \right) = \mu b|u|^{p-2} u \quad \text{in} \quad \Omega, \quad u = 0 \quad \text{on} \quad \partial\Omega,$$

in the k-weak sense.

2.3.3   A p-Laplacian problem in $\mathbb{R}^n$

Let Q be a positive continuous function on $\mathbb{R}^n$, $n \geq 1$, which is such that $\lim_{|\mathbf{x}|\to\infty} Q(\mathbf{x}) = \infty$, and $1 < p < \infty$. Let X be the completion of $C_0^\infty(\mathbb{R}^n)$ with respect to the norm $\|\cdot\|_X$ defined by

$$\|u\|_X^p := \int_{\mathbb{R}^n} \left\{ \sum_{j=1}^n |D_j u(\mathbf{x})|^p + Q(\mathbf{x})|u(\mathbf{x})|^p \right\} d\mathbf{x}.$$

Then X is continuously embedded in $\overset{0}{W}{}^{1}_{p}(\mathbb{R}^n)$, and hence in $L_p(\mathbb{R}^n)$. Furthermore, the embedding $E : X \to L_p(\mathbb{R}^n)$ is compact. For with $\varphi_t \in C_0^1([0, \infty))$ satisfying

$$\varphi_t(r) = \begin{cases} 1, & 0 \leq r < t, \\ 0, & r \geq t+1, \end{cases}$$

we have that $\varphi_t E : X \to L_p(B_{t+1})$ is compact since $\overset{0}{W}{}^1_p(\mathbb{R}^n) \hookrightarrow L_p(B_{t+1})$ is compact; here B_t stands for the open ball centre the origin and radius t in $\mathbb{R}^n$. Also, with $Y = L_p(\mathbb{R}^n)$,

$$\|(E - \varphi_t E)u\|_Y^p \le \int_{\mathbb{R}^n \setminus B_t} |u(\mathbf{x})|^p d\mathbf{x}$$
$$\le Q_t^{-1} \|u\|_X^p,$$

where $Q_t = \inf_{\mathbb{R}^n \setminus B_t} Q(\mathbf{x})$. The assumption that $Q_t(\mathbf{x}) \to \infty$ as $t \to \infty$ therefore means that E is the limit in norm of compact maps $\varphi_t E$ and hence is compact.

The duality maps of X and Y are

$$J_X(u) = -\Delta_p u + Q|u|^{p-2}u, \quad J_Y u = |u|^{p-2}u.$$

We may now apply our procedure to establish the existence of a sequence of j-eigenvalues and j-eigenvectors for the eigenvalue problem

$$-\Delta_p u + Q|u|^{p-2}u = \lambda |u|^{p-2}u, \quad u \in X, \tag{2.3.18}$$

in the k-weak sense.

In [15] a detailed analysis of the eigenvalue problem (2.3.18) is made in the case when Q is a radially symmetric C^1 function. The problem (2.3.18) then reduces to

$$-\mathcal{D}_p u + Q u^{(p-1)} = \lambda u^{(p-1)} \quad \text{in} \quad (0, \infty),$$
$$u'(0) = 0, \quad u \in L_p(0, \infty; r^{n-1} dr), \tag{2.3.19}$$

where $r = |\mathbf{x}|$, $L_p(0, 1; r^{n-1} dr)$ is the Lebesgue space with weight r^{n-1} and $\mathcal{D}_p$ is the radial p-Laplacian

$$\mathcal{D}_p u := r^{1-n} \left(r^{n-1} |u'|^{p-2} u' \right)'.$$

The following theorem is obtained.

Theorem 1. *Suppose that $Q \in C^1([0, \infty))$ is such that*

1. $Q(r) \ge \alpha r^\beta$ *for some* $\alpha > 0$, $\beta > \max\{\frac{p-n}{p-1}, 0\}$ *and* r *large enough;*

2. $\lim_{r \to \infty} \{\frac{Q'(r)}{Q(r)^{1+1/p}}\} = 0$.

Then (2.3.19) has a sequence of simple eigenvalues $\lambda_1 < \lambda_2 < \cdots$, with $\lambda_k \to \infty$ as $k \to \infty$, and no others. The eigenfunction u_k corresponding to λ_k has $k-1$ simple zeros in $(0, \infty)$. Moreover, if $z_j(k)$, $j = 1, 2, \ldots, k-1$, denote the zeros of u_k, then $0 < z_1(k+1) < z_1(k) < z_2(k+1) < z_2(k) < \cdots < z_{k-1}(k+1) < z_{k-1}(k) < z_k(k+1) < \infty$.

In the case $p = 2$, this is reminiscent of a result of Titchmarsh in [84], Section 5.9, p. 121, in which the condition $\lim_{r \to \infty} Q(r) = \infty$, and the assumptions that, at infinity, $Q'(r) > 0$, $Q''(r)$ is of one sign, and $Q'(r) = O(Q(r)^c)$ for $0 < c < 3/2$, are shown to imply that (2.3.19) is in the limit-point case at infinity and the spectrum of the associated self-adjoint operator is discrete. In fact, the single condition $\lim_{r\to\infty} Q(r) = \infty$ is shown in [13] to be sufficient for Theorem 1 to hold. Also in [13], it is proved that if $\lim_{r\to\infty} Q(r) = 0$, then (2.3.19) has a finite or countably infinite number of negative eigenvalues λ_k such that $\lambda_0 < \lambda_1 < \cdots < 0$, and no other eigenvalues. Moreover, the eigenfunction u_k has k simple zeros in $(0, \infty)$. An upper bound for the positive eigenvalues is obtained in [14] for decaying Q.

Potentials Q satisfying $Q(r) = -r^\alpha$ for large r with $\alpha > p/p - 1$, are also treated in [15]. In that case a boundary condition is added at infinity to (2.3.19) and the resulting problem is shown to have a sequence of eigenvalues $(\lambda_k)_{k\in\mathbb{Z}}$ with $\lim_{k\to\infty} \lambda_k = \infty$ and $\lim_{k\to-\infty} \lambda_k = -\infty$, and no others; each eigenfunction has an infinite number of zeros. When $p = 2$, the corresponding eigenvalue problem is in the limit-circle case at infinity.

2.3.4   The p-biharmonic operator

Theorem 2.2.5 can also be applied to the Dirichlet problem for the p-biharmonic operator $\Delta\left(|\Delta u|^{p-2}\Delta u\right)$ on a bounded open set $\Omega \subset \mathbb{R}^n$ for $1 < p < \infty$, namely,

$$\Delta\left(|\Delta u|^{p-2}\Delta u\right) = \mu|u|^{p-2}u \ \text{ in } \Omega, \quad u = |\nabla u| = 0 \ \text{ on } \partial\Omega. \tag{2.3.20}$$

In this case $Y = L_p(\Omega)$ and $X = \overset{0}{W}{}^2_p(\Omega)$, the completion of $C_0^\infty(\Omega)$ with respect to the norm

$$\|u\|_X := \left(\int_\Omega |\Delta u|^p d\mathbf{x}\right)^{1/p},$$

which is equivalent to the usual norm

$$\left(\sum_{|\alpha|\leq 2}\int_\Omega |D^\alpha u|^p dx\right)^{1/p}$$

in standard notation; see the remark following Corollary 7.11 in [49]. This means that the embedding $T = \mathrm{id} : X \to Y$ is compact. As in the previous examples, it is readily checked that X is reflexive and strictly convex, and that X^* is strictly convex. Also, the duality map J_X is given by the p-biharmonic operator:

$$J_X u = \Delta\left(|\Delta u|^{p-2}\Delta u\right).$$

Theorem 2.2.5 now establishes the existence, in the k-weak sense, of a countable family of j-eigenvectors u_k and corresponding j-eigenvalues μ_k^{-p} of (2.3.20), where

$$\mu_k = \|u_k\|_Y = \sup_{u\in X_k\setminus\{0\}}\left(\frac{\|u\|_Y}{\|u\|_X}\right).$$

As in Section 2.3.1, our approach also applies to the eigenvalue problem

$$\Delta\left(|\Delta u|^{p-2}\Delta u\right) = \mu|u|^{q-2}u, \text{ in } \Omega, \ u = |\nabla u| = 0 \text{ on } \partial\Omega \tag{2.3.21}$$

as long as $1 < p < n/2$, $p \leq q < np/(n-2p)$. For then $X = \overset{0}{W^2_p}(\Omega)$ is compactly embedded in $L_q(\Omega)$ and all the requirements of Theorem 2.2.5 are met with $T = \mathrm{id}$ the embedding. If $p = 2$ the j-eigenvectors form an orthonormal basis for X, while if $q = 2$, they form an orthonormal basis of Y.

2.3.5 Sturm–Liouville theory for the p-Laplacian

We shall now follow [9] and discuss a Sturm–Liouville type boundary-value problem associated with the equation

$$-\left(s^{-(p-1)}|y'|^{p-2}y'\right)' = (\gamma r - q)|y^{p-2}|y \tag{2.3.22}$$

on (a, b), $-\infty < a < b < \infty$, where q, r, s are integrable functions on (a, b), r, s are positive a.e., and $\gamma \in \mathbb{R}$ is the eigenvalue parameter. We shall use the notation

$$u_{(p)} := |u|^{p-2}u.$$

The analysis in [9] is based on the following Prüfer-type transformation which was used by Elbert in [44]:

$$\rho\sin_p\theta = y, \quad \rho\cos_p\theta = \frac{fy'}{s}, \tag{2.3.23}$$

where $\sin_p, \cos_p$ are the functions defined in Chapter 1, Section 1.3, and, in our case, f is a positive constant (possibly depending on γ), which will be chosen later. From (2.3.22) and the fact that

$$(\cot_p\theta)_{(p)} = \left(\frac{fy'}{sy}\right)_{(p)}, \tag{2.3.24}$$

where $\cot_p\theta = \cos_p\theta / \sin_p\theta$, it is readily shown that

$$(1 + |\cot_p\theta|^p)\,\theta' = f^{p-1}(1-p)^{-1}(\gamma r - q) + \frac{s}{f}|\sin_p\theta|^p. \tag{2.3.25}$$

On using (2.3.22) and the identity $|\sin_p\theta|^p + |\cos_p\theta|^p = 1$, we obtain

$$\theta' = f^{p-1}(1-p)^{-1}(\gamma r - q)|\sin_p\theta|^p + \frac{s}{f}|\cos_p\theta|^p. \tag{2.3.26}$$

Furthermore, from (2.3.23),

$$\rho' = \rho\cot_p\theta\left(\frac{s}{f} - \theta'\right). \tag{2.3.27}$$

Denote the right-hand side of (2.3.26) by $G(\cdot, \gamma, \theta)$. On using the mean-value theorem, we obtain

$$|G(\cdot, \gamma, \theta) - G(\cdot, \gamma, \varphi)| = O\left(\gamma r f^{p-1} + s f^{-1}\right) |\theta - \varphi|.$$

Hence, the initial-value problem for θ has a unique solution which satisfies

$$\theta(x, \lambda) - \theta(a, \lambda) = \int_a^x G(t, \gamma, \theta) dt.$$

Setting $f = 1$, and with the initial value of θ at a independent of γ, we see that θ is continuous in γ and pointwise increasing in γ. Since the equation (2.3.27) is linear in ρ with an integrable coefficient, it follows that the initial-value problem for ρ, θ, and thus for y, y', is uniquely solvable.

Let φ satisfy (2.3.26) with $f = (p-1)^{1/p} \gamma^{-1/p}$ and $|\theta(a) - \varphi(a)| < \pi_p/2$. Then

$$\varphi' = \left(\frac{\gamma}{p-1}\right)^{1/p} \{r |\sin_p \varphi|^p + s |\cos_p \varphi|^p\}$$

$$\geq \left(\frac{\gamma}{p-1}\right)^{1/p} \min\{r, s\}.$$

Therefore $\varphi(b, \gamma) \to \infty$ as $\gamma \to \infty$. Since (y, y') and $(y, \gamma^{1/p} y')$ lie in the same quadrant, we have that for all $x \in [a, b]$,

$$|\theta(x, \gamma) - \varphi(x, \gamma)| < \pi_p/2$$

and so $\theta(b, \lambda) \to \infty$ as $\gamma \to \infty$.

Consider now the boundary-value problem consisting of (2.3.22) and the boundary conditions

$$\frac{y'(a)}{s(a)y(a)} = \cot_p \theta_a, \tag{2.3.28}$$

$$\frac{y'(b)}{s(b)y(b)} = \cot_p \theta_b, \tag{2.3.29}$$

where θ_a and θ_b are independent of γ; the Dirichlet conditions $y(a) = 0$, $y(b) = 0$ correspond to $\theta_a = 0, \theta_b = \pi_p$ respectively. This corresponds to the eigenvalue problem for θ with the boundary conditions $\theta(a) = \theta_a, \theta(b) = \theta_b + n\pi_p$ for some integer n. The following theorem is a consequence of the above discussion. Note that since (2.3.22) is homogeneous, eigenfunctions are unique only up to constant multiples.

Theorem 2.3.1. *The eigenvalue problem* (2.3.22), (2.3.28), (2.3.29) *has a countable infinity of simple eigenvalues* $\gamma_1 < \gamma_2 < \cdots$ *which accumulate at infinity. The eigenfunction* y_n *corresponding to* γ_n, *and its derivative* y'_n, *have precisely* $(n-1)$ *zeros in* (a, b) *and the zeros of* y_n *and* y'_n *are interlaced.*

The one-dimensional case of (2.3.9) is of the form

$$- \left(u'_{(p)} \right)' = \gamma u_{(p)}. \tag{2.3.30}$$

with Dirichlet boundary conditions

$$u(a) = u(b) = 0. \tag{2.3.31}$$

It is shown in [28] that when $a = 0, b = 1$, the eigenvalues and corresponding eigenvectors of (2.3.30)–(2.3.31) are

$$\gamma_n = (p-1)(n\pi_p)^p, \quad u_n(t) = \sin_p(n\pi_p t), \quad (n \in \mathbb{N}), \tag{2.3.32}$$

where

$$\pi_p = \frac{2\pi}{p \sin_p(\pi/p)}.$$

2.4 Lusternik–Schnirelmann critical levels

2.4.1 Comparison of eigenvalues

For simplicity, set $s = 1$ in the last subsection and define

$$W := \{ y \in W_p^1(a, b) : y \text{ satisfies } (2.3.28) \text{ and } (2.3.29) \}. \tag{2.4.1}$$

Then

$$\int_a^b \{ -[y'_{(p)}]' + q y_{(p)} \} \bar{y} dt = \int_a^b \{ |y'|^p + q|y|^p \} dt$$

$$=: I(y).$$

It follows that the eigenfunctions and eigenvalues of (2.3.22), (2.3.28) and (2.3.29) correspond respectively to critical points and critical levels of the functional $I(y)$ on

$$S := \{ y \in W : \int_a^b r|y|^p dt = 1 \}.$$

In order to define the kth critical level of the functional $I(\cdot)$ on S in the Lusternik–Schnirelmann theory, we need some additional terminology and notation. Let $\mathcal{A}$ be the set of all compact subsets A of $S \setminus \{0\}$ which are symmetric, i.e., $A = -A$. The **Krasnoselski genus** $\gamma(A)$ of A is given by

$$\gamma(A) := \inf\{ j \in \mathbb{N} : \exists \text{ a continuous, odd function } f : A \to \mathbb{R}^j \setminus \{0\} \}.$$

The kth **Lusternik–Schnirelmann** (L-S for short) critical level of $I(\cdot)$ on S is defined by

$$\tau_k = \inf_{A \in \mathcal{F}_k} \sup_{u \in A} I(u) \tag{2.4.2}$$

where
$$\mathcal{F}_k := \{A \in \mathcal{A} : \gamma(A) \geq k\},$$

The following theorem is proved in [9].

Theorem 2.4.1. *For all $n \in \mathbb{N}$, the nth eigenvalue of (2.3.22), (2.3.28) and (2.3.29) coincides with the nth L-S critical level of $I(\cdot)$ on S, i.e.,*

$$\gamma_n = \tau_n \quad (n \in \mathbb{N}).$$

Proof. It is shown in the proof of Theorem 5.1 in [9] that since, by Theorem 2.3.1, all the eigenvalues of (2.3.22), (2.3.28) and (2.3.29) are simple and given by the $\gamma_n, n \in \mathbb{N}$, then $\tau_n \geq \gamma_n$. It is therefore sufficient to prove the reverse inequality.

Let the eigenfunction y_n associated with γ_n be normalised:

$$\int_a^b r|y_n|^p dt = 1.$$

Let $\mathcal{I}_i \subset (a, b), i = 1, 2, \ldots, n$, be the subintervals on which y_n is of definite sign and set $w_n := y_n \chi_i$, where χ_i is the characteristic function of $\mathcal{I}_i$. Then

$$\mathcal{A} := \left\{ y = \sum_{i=1}^n a_i w_i : \int_a^b r|y|^p dt = \sum_{i=1}^n |a_i|^p \int_{\mathcal{I}_i} r|w_i|^p dt = 1 \right\}$$

is symmetric and homeomorphic to $\mathbb{S}^{n-1}$ and hence $\mathcal{A} \in \mathcal{F}_n$. Furthermore, since the subintervals $\mathcal{I}_i$ are non-overlapping, it follows from (2.3.22) that

$$I(y) = \sum_{i=1}^n \left(|a_i|^p \int_{\mathcal{I}_i} \{|w_i'|^p + q|w_i|^p\}dt \right)$$

$$= \gamma_n \sum_{i=1}^n |a_i|^p \int_{\mathcal{I}_i} r|w_i|^p dt = \gamma_n.$$

Hence $\gamma_n \geq \tau_n$ and the proof is complete. $\qquad\square$

In [61] the coincidence of eigenvalues and L-S critical levels was investigated for the homogeneous p-Laplace equation

$$-\Delta_p u := -\sum_{j=1}^n D_j \left(|D_j u|^{p-2} D_j u \right) = \gamma |u|^{p-2} u \qquad (2.4.3)$$

on a bounded smooth domain $\Omega \subset \mathbb{R}^n$ subject to a variety of conditions on the boundary $\partial\Omega$, in particular the following:

- Dirichlet:
$$u = 0 \text{ on } \partial\Omega;$$

- Neumann:

$$\frac{\partial u}{\partial \nu} = 0 \ \text{ on } \ \partial \Omega,$$

where $\partial u / \partial \nu$ denotes the normal derivative;

- Robin:

$$|\nabla u|^{p-2}\frac{\partial u}{\partial \nu} + \beta |u|^{p-2}u = 0 \ \text{ on } \ \partial \Omega,$$

for some $\beta \in [0, \infty)$.

The L-S sequence (τ_k) is now defined by

$$\tau_k := \inf_{A \in \mathcal{F}_k} \sup_{u \in A} G(u), \quad G(u) := \left(\int_\Omega (|\nabla u|^p + |u|^p) \, d\mathbf{x} + \beta \int_{\partial \Omega} |u|^p dx \right),$$

where $\beta = 0$ for the Dirichlet and Neumann problems, and $\mathcal{F}_k$ is the set of all compact symmetric subsets A of the level set $\{u \in W_p^1(\Omega) : \int_\Omega |u|^p d\mathbf{x} = 1\}$ which are such that $G(u) > 0$ on A and the genus $\gamma(A) \geq k$.

It is proved in [61], *inter alia*, that for each of the Dirichlet, Neumann and Robin problems, the $\tau_k, k \in \mathbb{N}$, are eigenvalues and satisfy the following:

- τ_1 is simple and only eigenfunctions associated with τ_1 do not change sign;
- τ_2 is the second eigenvalue.

We refer to [61], Remark 5.5, for other references which are relevant to this result.

It is still unknown if there are any eigenvalues of the above problems which do not belong to the L-S sequence (τ_k). In general, such eigenvalues are known to exist. For instance, let ε be a fixed positive number, q a real-valued, continuous $(2\pi_p)$-periodic function on $\mathbb{R}$ and consider the eigenvalue problem

$$-((u')_{(p)})'(x) + \varepsilon q(x)u_{(p)}(x) = \gamma u_{(p)}(x), \quad x \in \mathbb{R}, \tag{2.4.4}$$

with u required to be $(2\pi_p)$-periodic in $\mathbb{R}$; γ is the eigenvalue parameter. It is proved in [10], Theorem 5.1, that there exists a sequence (γ_k^ε) of L-S eigenvalues which tends to infinity, and is such that γ_1^ε is simple, the corresponding eigenfunction does not change sign and for $n \geq 1$,

$$\gamma_{2n-1}^\varepsilon \leq \gamma_{2n}^\varepsilon < \gamma_{2n+1}^\varepsilon \leq \gamma_{2n+2}^\varepsilon.$$

Furthermore, in [10], Theorems 4.5 and 5.1, it is proved that, for any integers $m, n \geq 1$ and for any sufficiently small ε, there exists a $(2\pi_p)$- periodic function $q \in C^1(\mathbb{R})$ such that $\gamma_{2n-1}^\varepsilon < \gamma_{2n}^\varepsilon$ and the open interval $(\gamma_{2n-1}^\varepsilon, \gamma_{2n}^\varepsilon)$ contains at least m "non-variational" eigenvalues, i.e., they do not belong to the L-S sequence. In [30], Theorem 1.1, it is proved that some of these exceptional eigenvalues do, nevertheless, admit a "local" variational characterisation.

2.4.2   A Hardy-type operator

Let $-\infty < a < t < b < \infty$, $1 < p < \infty$, and let T be the integral operator defined on $L_p(a, b)$ by

$$Tf(t) := v(t) \int_a^t f(s)u(s)ds, \qquad (2.4.5)$$

where u, v are real functions satisfying $u \in L_{p'}(a, b)$ and $v \in L_p(a, b)$, where $p' = p/(p-1)$; the Hardy operator is, of course, the special case $u = v = 1$. Under the conditions on u and v, T is a compact linear map of $L_p(a, b)$ into itself. It is bounded since, by Hölder's inequality,

$$\|Tf\|_p^p \leq \int_a^b \left\{ |v(t)| \left(\int_a^t |u(s)|^{p'} ds \right)^{1/p'} \right\}^p dt \int_a^b |f(s)|^p ds$$

$$\leq (\|u\|_{p'} \|v\|_p)^p \|f\|_p^p,$$

and so $\|T\| \leq \|u\|_{p'} \|v\|_p$. To prove that T is compact, set $Gf(t) = \int_a^t u(s)f(s)ds$. Then

$$|Gf(t) - Gf(x)| \leq \left(\int_x^t |u(s)|^{p'} ds \right)^{1/p'} \|f\|_p.$$

Let (f_n) be a sequence in the closed unit ball of $L_p(a, b)$. Then (Gf_n) is uniformly bounded and equicontinuous on $[a, b]$, and so, by the Arzelà–Ascoli theorem, it contains a subsequence which converges uniformly on (a, b). Since $\|Tf\|_p = \|vGf\|_p$, the corresponding subsequence of (Tf_n) converges in $L_p(a, b)$. Thus T maps the closed unit ball in $L_p(a, b)$ into a relatively compact subset of $L_p(a, b)$, which proves the asserted compactness of T.

The adjoint T^* of T is

$$T^*h(t) = u(t) \int_t^b h(s)v(s)ds; \qquad (2.4.6)$$

it is a compact linear map of $L_{p'}(a, b)$ into itself.

The results of this section come mainly from [7]. We shall assume that u and v do not vanish on (a, b), but in [7] they are allowed to vanish on sets of positive measure.

Consider equation (2.2.7),

$$T^*J_Y Tf = \nu J_X f, \quad \nu = \|T\| \mu_Y(\|T\|), \qquad (2.4.7)$$

which comes from the Euler equation for maximising $\|Tf\|_Y / \|f\|_X$. Then, with $X = Y = L_p(a, b)$ and gauge function $\mu_X(t) = t^{p-1}$, we have that the duality map J_X is given by

$$J_X(f) = |f|^{p-2} f.$$

On substituting this in (2.4.7), and setting $g(t) = (Tf)(t)/v(t)$, we obtain

$$u(t) \int_t^b g|g|^{p-2}|v|^p ds = \nu|u|^{-p}g'|g'|^{p-2}, \tag{2.4.8}$$

where $\nu = \|T\|^p$. With $\gamma := \nu^{-1/p}$, this gives a nonlinear eigenvalue problem of Sturm–Liouville type:

$$-\left(|u|^{-p}g'|g'|^{p-2}\right)' = \gamma|v|^p g|g|^{p-2}$$
$$g(a) = 0, \quad g'(b) = 0. \tag{2.4.9}$$

This is the special case $s = |u|^{p/(p-1)}, r = |v|^p, q = 0, \theta_a = 0, \theta_b = \pi_p/2$ of (2.3.22), (2.3.28) and (2.3.29). The following result is therefore obtained from Theorem 2.3.1; see also [7], Section 2.

Theorem 2.4.2. *There is a sequence (γ_n) of simple eigenvalues of (2.4.9) tending to infinity. The corresponding eigenfunctions g_n and their derivatives g_n' have precisely $n - 1$ zeros in (a, b), and the zeros of g_n and g_n' are interlaced.*

A notable result from [7] is that with $(\gamma_n)_{n \in \mathbb{N}}$ the eigenvalues in Theorem 2.4.2 and $\mu_n := \gamma_n^{-1}$, then for all $n \in \mathbb{N}$, the μ_n coincide with the approximation numbers $a_n(T)$ and Bernstein numbers $b_n(T)$ of the operator (2.4.5). Before establishing this, we first recall that for any bounded linear operator $T : X \to Y$,

$$a_n(T) = \inf\{\|T - P\| : P \in B(X, Y), \text{ rank } P < n\}$$

and

$$b_n(T) = \sup_{\dim B \geq n} \inf \left\{ \frac{\|Tf\|_Y}{\|f\|_X} : f \in B, f \neq 0 \right\},$$

where B is a subspace of X. Since the approximation numbers are the largest s-numbers (see Chapter 1, Section 4), we have for any $T \in B(X, Y)$,

$$b_n(T) \leq a_n(T), n \in \mathbb{N}. \tag{2.4.10}$$

Theorem 2.4.3. *For the operator T in (2.4.5), $\mu_n = a_n(T) = b_n(T)$.*

Proof. In view of (2.4.10), the result will follow if we prove that $a_n(T) \leq \mu_n$ and $\mu_n \leq b_n(T)$. The proof hinges on two facts: that the first eigenvalue γ_1 of (2.4.9) is $\|T\|^{-1}$ (and so $\mu_1 = \|T\|$), and the first eigenfunction has no zeros in (a, b). With reference to Theorem 2.4.2, denote the zeros of the eigenfunctions g_n corresponding to $\mu_n = \gamma_n^{-1}$ by $a = s_0, s_1, \ldots, s_{n-1}$ and those of g_n' by $d_1, d_2, \ldots, d_n = b$. By the interlacing property, $s_{j-1} < d_j < s_j$, $j = 1, 2, \ldots, n - 1$. Define $P : L_p(a, b) \to L_p(a, b)$ by

$$Pf(x) := \sum_{j=1}^{n-1} g(s_j)v(x)\chi_j(x),$$

where $g(x) = (Tf/v)(x) = \int_a^x fuds$ and χ_j is the characteristic function of the interval $I_j = (d_j, d_{j+1})$. Thus $a_n(T) \le \|T - P\|$. The operator $T - P$ is the direct sum of operators T_j acting on $L_p(I_j), j = 0, 1, \ldots, n - 1$, with $I_0 = (a, d_1)$. For $j \ge 1$, each T_j is the direct sum of operators $T_j^{(1)}, T_j^{(2)}$, similar in form to T_j, and acting in $L_p(d_j, s_j), L_p(s_j, d_{j+1})$, respectively: to be specific,

$$T_j^{(1)}f(x) = -v(x)\int_x^{s_j} fuds, \quad x \in (d_j, s_j),$$

$$T_j^{(2)}f(x) = v(x)\int_{s_j}^x fuds, \quad x \in (s_j, d_{j+1}).$$

These are Hardy-type operators which lead to eigenvalue problems like (2.4.9) but with the boundary conditions at d_j and s_j interchanged in the case of $T_j^{(1)}/v$. The restrictions of g_n to (d_j, s_j) and (s_j, d_{j+1}) satisfy these eigenvalue problems with $\gamma = \mu_n^{-1}$ and have no interior zeros in each case. Thus $\mu_n = \|T_j^{(1)}\| = \|T_j^{(2)}\|$ and hence $\mu_n = \|T_j\|$. The same applies to the interval $I_0 = (a, d_1)$ and the corresponding operator T_0. Then, for any $\varphi \in L_p(a, b)$ and setting $\varphi_j = \varphi\chi_j$, we have

$$\|T\varphi - P\varphi\|_p^p = \sum_{j=0}^{n-1}\|T_j\varphi_j\|_p^p \le \sum_{j=0}^{n-1}\|T_j\|^p\|\varphi_j\|_p^p$$

$$= \mu_n^p \sum_{j=0}^{n-1}\|\varphi_j\|_p^p = \mu_n^p\|\varphi\|^p.$$

Therefore $a_n(T) \le \|T - P\| \le \mu_n$.

The next, and final, step, is to prove that $\mu_n \le b_n(T)$. We shall construct a subspace B of $L_p(a, b)$ of dimension n, such that $\mu_n\|f\|_p \le \|Tf\|_p$ for all $f \in B$. Let $f = g_n'/u$, and put $f_j = f\chi_j, j = 0, 1, 2, \ldots, n - 1$. The linear hull B of $f_1, \ldots, f_n$ is therefore n-dimensional. Furthermore, we have $\|T_j f_j\|_p = \mu_n\|f_j\|_p$. Thus, if $\varphi = \sum_{j=0}^{n-1} a_j f_j, a_j \in \mathbb{R}$,

$$\mu_n^p\|\varphi\|_p^p = \mu_n^p\sum_{j=0}^{n-1}|a_j|^p\|f_j\|_p^p = \sum_{j=0}^{n-1}|a_j|^p\|T_j f_j\|_p^p$$

$$= \sum_{j=0}^{n-1}\|T_j(a_j f_j)\|_p^p = \|T\varphi\|_p^p.$$

The proof is therefore complete.   □

In the case of the Hardy operator, that is, when $u = v = 1$, on $(0, 1)$, Theorem 2.4.3 is contained in [38], Theorem 5.1, which proves that the μ_n are equal to all the "so-called strict" s-numbers and thus, in particular, to the approximation and Bernstein numbers; see Section 1.4.1 for the definition and properties of s-numbers. From [28] it follows that the eigenvalues and eigenfunctions of (2.4.9) with $u = v = 1, a = 0, b = 1$ are

$$\gamma_n = \{(p-1)[(n-1/2)\pi_p]\}, \quad u_n(t) = \frac{1}{(n-1/2)\pi_p} \sin_p[(n-1/2)\pi_p t], \quad n \in \mathbb{N}.$$

$$(2.4.11)$$

The relation between these eigenvalues and their counterparts obtained by the Lusternik–Schnirelmann method is determined in [38], Theorem 5.2. The Lusternik–Schnirelmann kth *critical level* for T is now

$$c_k(T) = \sup_{F \in C_k^0} \left\{ \min_{u \in F \subset L_p(0,1), u \neq 0} \frac{\|Tu\|_p}{\|u\|_p} \right\},$$

where C_k^0 is the set of all compact symmetric subsets F of $L_p(0,1) \setminus \{0\}$ such that the genus of F,

$$\gamma(F) := \{l \in \mathbb{N} : \text{there exists } f \in C(F, \mathbb{R}^l \setminus \{0\}), \ f(x) = f(-x)\} \geq k.$$

Theorem 2.4.4. *Let $1 < p < \infty, n \in \mathbb{N}$ and let T be the Hardy operator acting from $L_p(0,1)$ to itself. Then*

$$c_n(T) = \mu_n = \gamma_n^{-1} = \{(p-1)(n-1/2)\pi_p\}^{-1}.$$

where γ_n is the nth eigenvalue of (2.4.9). The critical level is attained only for functions which are derivatives of the nth eigenfunction

$$u_n(t) = \frac{1}{(n-1/2)\pi_p} \sin_p[(n-1/2)\pi_p t].$$

Proof. It follows from [61], Theorem 2.5, that $c_k^{-1}(T)$ is an eigenvalue of (2.4.9) and so $(c_k^{-1}(T))_{k \in \mathbb{N}} \subset (\gamma_n)_{n \in \mathbb{N}}$. This implies that $c_k^{-1}(T) \geq \gamma_k$.

Consider the nth eigenfunction u_n of (2.4.9), corresponding to the nth eigenvalue γ_n. Set $w_i(x) = u_n(x)\chi_{I_i}(x)$, where $I_i = (a_{i-1}, a_i]$, with $a_0 = 0, a_i = i(n-1/2)^{-1}$ for $i = 1, 2, \ldots, n-1$, and $a_n = 1$. Thus the I_i are non-overlapping intervals which cover $(0,1)$. Then the set $C_k := \text{sp}\{w_1, w_2, \ldots, w_n\} \cap (L_p(0,1) \setminus \{0\})$ has genus n and $C_k \subset C_k^0$. From this it follows that $c_k^{-1}(T) \leq \gamma_n$ and hence the proof is complete. $\qquad \square$

It isn't at all clear if the j-eigenvalues and j-eigenfunctions of T are related to its eigenvalues and eigenfunctions, excepting, of course, the first ones, which coincide. The semi-orthogonality property of the j-eigenfunctions and the fact that they can form a basis of X in the general case of $T : X \to Y$ and Banach spaces X, Y, suggest that they have an important role to play and are worthy of investigation. That the eigenfunctions are not in general semi-orthogonal is suggested by numerical computations mentioned in [38] and relating to the Hardy operator T. Since the eigenvalues of T are simple in view of Theorem 2.4.2, it follows that its j-eigenvalues are not eigenvalues.

2.5   Further consequences of the boundedness of $(S_n)_{n\in\mathbb{N}}$

We proved in Corollary 2.2.23 that if the kernel of T is trivial and the hypothesis that the sequence $(S_n)_{n\in\mathbb{N}}$ is bounded is made, then $(x_n), (\xi_n)$ are bases of X, X^*, respectively, and the action of T is given by a series. In this section we derive more consequences of this hypothesis. Throughout this section we shall therefore assume, in addition to the standing assumptions on X, Y and T, (namely, that X, Y are reflexive, strictly convex Banach spaces with strictly convex duals, and T is a compact linear map of X into Y), that $(S_n)_{n\in\mathbb{N}}$ is bounded and that T has trivial kernel.

To obtain further properties of the ξ_n we shall also suppose that X is uniformly convex and uniformly smooth; X^* then has the same properties. By Corollary 2.2.23, the functionals ξ_n form a basis of X^*, so that each $x^* \in X^*$ has a unique representation of the form $x^* = \sum_{n=1}^{\infty} a_n\xi_n$. Define projections $U_N : X^* \to X^*$ by $U_N x^* = \sum_{n=1}^{N} a_n\xi_n$ ($N \in \mathbb{N}$), and let b be the basis constant of the basis (ξ_n): $b = \sup_N \|U_N\| < \infty$. Let δ be the modulus of convexity of X^*. Examination of the proofs of Proposition 1.2.17 (applied to X^* and its basis (ξ_n)) and Proposition 1.2.18 shows that, with $\varepsilon \in (0, 1/b)$, $\lambda = 2(1 - \delta(\varepsilon))$, $r \in (1, \log_\lambda 2)$ and $q = r' = r/(r-1)$, there exists $K > 0$ such that for all $x \in X$,

$$\left(\sum_{n=1}^{\infty} |\xi_n(x)|^q\right)^{1/q} \leq K \|x\|_X .$$

Thus $(\xi_n(x)) \in l_q$ for each $x \in X$. To obtain a better idea of the restriction on q, observe that

$$\log_\lambda 2 = \frac{\log 2}{\log 2 + \log(1 - \delta(\varepsilon))} = \left(1 + \frac{\log(1 - \delta(\varepsilon))}{\log 2}\right)^{-1}$$
$$> 1 - \frac{\log(1 - \delta(\varepsilon))}{\log 2} > 1 + \frac{\delta(\varepsilon)}{\log 2}.$$

The choice of r so that

$$1 < r < 1 + \frac{\delta(\varepsilon)}{\log 2}$$

is thus a valid one, and the corresponding $q = r'$ satisfies

$$0 < \frac{1}{q} < 1 - \frac{1}{1 + \frac{\delta(\varepsilon)}{\log 2}} = \frac{\delta(\varepsilon)}{\log 2 + \delta(\varepsilon)},$$

so that

$$q \in \left(1 + \frac{\log 2}{\delta(\varepsilon)}, \infty\right).$$

Since ε may be chosen arbitrarily close to $1/b$, we have

Theorem 2.5.1. *Let X be uniformly convex and uniformly smooth. Then for every $x \in X$, the sequence $(\xi_n(x))$ belongs to l_q if $q > 1 + \frac{\log 2}{\delta(1/b)}$, where b is the basis constant corresponding to the basis (ξ_n) of X^*.*

Greater precision is possible on specialisation of X. For example, we have

Corollary 2.5.2. *Suppose that X is an L_p space, where $p \in (1, \infty)$, and assume that T has trivial kernel. Then for every $x \in X$, the sequence $(\xi_n(x))$ belongs to l_q if $q > 1 + p'(2b)^{p'} \log 2$ when $1 < p \leq 2$, and $q > 1 + 8b^2(p - 1) \log 2$ when $2 < p < \infty$. Here b is the basis constant corresponding to the basis (ξ_n) of $L_{p'}$.*

Proof. Corollaries 1.1.9 and 1.1.11 show that the modulus of convexity δ of $L_{p'}$ satisfies

$$
\delta(\varepsilon) \geq \begin{cases} (\varepsilon/2)^{p'}/p', & 1 < p \leq 2, \\ (p' - 1)\varepsilon^2/8, & 2 < p < \infty. \end{cases}
$$

From Theorem 2.5.1 it follows that if $1 < p \leq 2$, then q may be chosen arbitrarily in the interval $\left(1 + p'(2b)^{p'} \log 2, \infty\right)$, while if $2 < p < \infty$, the interval $\left(1 + \frac{8b^2 \log 2}{(p'-1)}, \infty\right)$ is appropriate. The result follows. $\qquad\square$

This result is somewhat reminiscent of the much more precise information given by the Hausdorff–Young theorem about the Fourier coefficients of a function in $L_p(0, 2\pi)$, where $1 < p \leq 2$, namely that they form a sequence in $l_{p'}$. To sharpen the corollary an upper bound for the basis constant is needed.

The representation of T in Corollary 2.2.23 can be written in tensor product notation as

$$
T = \sum_{n=1}^{\infty} \lambda_n \xi_n \otimes y_n,
$$

where $y_n = Tx_n/\|Tx_n\|_X$. Given any $r \in (0, 1]$, the map T is said to be *r*-**nuclear** (see [59], 1.d) if there are sequences (x_n^*) and (z_n) in X^* and Y respectively such that

$$
T = \sum_{n=1}^{\infty} x_n^* \otimes z_n, \quad \text{with} \quad \sum_{n=1}^{\infty} \|x_n^*\|_{X^*}^r \|z_n\|_Y^r < \infty;
$$

the *r*-**nuclear norm** of T is

$$
\nu_r(T) := \inf \left(\sum_{n=1}^{\infty} \|x_n^*\|_{X^*}^r \|z_n\|_Y^r \right)^{1/r},
$$

where the infimum is taken over all representations of T of the form

$$
T = \sum_{n=1}^{\infty} x_n^* \otimes z_n.
$$

The nuclear maps are those with $r = 1$. In view of (1.2.2) in Chapter 2, we have established

Proposition 2.5.3. *If* $(\lambda_n)_{n \in \mathbb{N}} \in l_r$, *then* T *is r-nuclear and*

$$\nu_r(T) \le 2b \, \|(\lambda_n)\|_r \,,$$

where $b = \sup_{n \in \mathbb{N}} \|S_n\|$ *is the basis constant corresponding to the basis* (x_n) *of* X *and* $\|\cdot\|_r$ *denotes the* l_r *norm.*

From the definition of the Gelfand numbers $c_n(T)$ of T it is immediate that for all $n \in \mathbb{N}$,

$$c_n(T) \le \lambda_n,$$

where $\lambda_n = \|T_n\|$, $T_n = T \upharpoonright_{X_n}$. Let $p \in (0, \infty)$, $q \in (0, \infty]$ and denote by $l_{p,q}$ the usual Lorentz sequence space: this is the space of all sequences of scalars (f_n) such that $\|(f_n)\|_{p,q} < \infty$, where

$$\|(f_n)\|_{p,q} = \begin{cases} \left(\sum_{n=1}^{\infty} |f_n^*|^q \, n^{q/p-1}\right)^{1/q}, & q < \infty, \\ \sup_{n \in \mathbb{N}} n^{1/p} \, |f_n^*|, & q = \infty. \end{cases}$$

Here (f_n^*) is the non-increasing rearrangement of (f_n). Now suppose that $X = Y$ and denote the eigenvalues of the compact map $T : X \to X$ (if any) by $\tilde{\lambda}_n$. If $(\lambda_n) \in l_{p,q}$, then by Weyl's inequality in Lorentz form (see [59], p. 85),

$$\left\|(\tilde{\lambda}_n)\right\|_{p,q} \le c_{p,q} \, \|(w_n(T))\|_{p,q} \le c_{p,q} \, \|(c_n(T))\|_{p,q} \le c_{p,q} \, \|(\lambda_n)\|_{p,q} \,,$$

where

$$c_{p,q} = \begin{cases} 2^{1/p} \sqrt{e}, & p \ge q, \\ (2e)^{1/p+1/2}, & p < q. \end{cases}$$

Thus $(\lambda_n) \in l_{p,q}$ implies that $(\tilde{\lambda}_n) \in l_{p,q}$. Another connection with eigenvalues is given by the result (see [59], 2.d.6) that

$$\left|\tilde{\lambda}_n(T)\right| = \lim_{k \to \infty} \left(c_n\left(T^k\right)\right)^{1/k} \quad (n \in \mathbb{N}).$$

In view of this

$$\left|\tilde{\lambda}_n(T)\right| \le \liminf_{k \to \infty} \left(\lambda_n\left(T^k\right)\right)^{1/k}.$$

Moreover, we have

Proposition 2.5.4. *Let*

$$\mathfrak{L}_p^{(\lambda)} := \{T \in B(X) : (\lambda_n(T)) \in l_p\},$$

and define $\mathfrak{L}_p^{(c)}$ *in a similar way with* $c_n(T)$ *replacing* $\lambda_n(T)$. *Then*

$$\mathfrak{L}_p^{(\lambda)} \subset \mathfrak{L}_p^{(c)};$$

that is, the maps T *for which* $(\lambda_n) \in l_p$ *form a subset of the ideal* $\mathfrak{L}_p^{(c)}$ *formed by the Gelfand numbers, an ideal that is just the Schatten p-class when* X *is a Hilbert space.*

We do not know, however, whether or not this subset is a linear space.

The entropy numbers of an r-nuclear map $S : L_q \to L_q$ (where $1 < q < \infty$) have the property that $(e_n(S)) \in l_{s,r}$, where $1/s = 1/r - |1/2 - 1/q|$ (see [59]), so that by Carl's inequality (see (1.4.8))

$$|\mu_n(S)| \leq \sqrt{2} e_n(S), \quad n \in \mathbb{N},$$

for eigenvalues $\mu_n(S)$ and entropy numbers $e_n(S)$ of S, we have $(\tilde{\lambda}_n(S)) \in l_{s,r}$. It is known (see [18]) that in this case $((e_n S))$ does not, in general, belong to l_{s,r_0} if $r_0 < r$. If $q \neq 2$, then $s > r$ and l_r is a proper subset of $l_{s,t}$ for all $t \in (1, \infty)$. Hence those maps $S : L_q \to L_q$ for which $(\lambda_n(S)) \in l_r$ form a proper subset of the family of all r-nuclear maps $S : L_q \to L_q$: this underlines the strength of the hypothesis that $(\lambda_n(S)) \in l_r$.

Finally, let X be uniformly convex and uniformly smooth, and suppose that $(\lambda_n(T)) \in l_{p,\infty}$ for some $p \in (0, \infty)$. Then $(c_n(T)) \in l_{p,\infty}$. We also know from Theorem 2.5.1 that there exists $q > 1$ such that for each $x \in X$, $(\xi_n(x)) \in l_q$. Thus we have the following estimate for the approximation numbers of T.

Proposition 2.5.5. *Let X be uniformly convex and uniformly smooth. Then for all $N \in \mathbb{N}$,*

$$a_N(T) \leq C_1 N^{-1/p+1/q'}.$$

Proof. Since $(\lambda_n(T)) \in l_{p,\infty}$ and $(\xi_n(x)) \in l_q$, it follows that

$$a_N(T) \leq \sup_{x \in B_X} \left\| Tx - \sum_{n=1}^{N} \lambda_n \xi_n(x) y_n \right\|_X = \sup_{x \in B_X} \left\| \sum_{n=N+1}^{\infty} \lambda_n \xi_n(x) y_n \right\|_X$$

$$\leq \sup_{x \in B_X} \sum_{n=N+1}^{\infty} |\lambda_n \xi_n(x)|$$

$$\leq \sup_{x \in B_X} \left(\sum_{n=N+1}^{\infty} |\xi_n(x)|^q \right)^{1/q} \left(\sum_{n=N+1}^{\infty} |\lambda_n|^{q'} \right)^{1/q'}$$

$$\leq C \left(\sum_{n=N+1}^{\infty} n^{-q'/p} \right)^{1/q'} \leq C_1 N^{-1/p+1/q'}. \qquad \square$$

If $q \leq 2$, this is better than the estimate obtainable from the general inequality

$$a_n(T) \leq 2\sqrt{n} c_n(T)$$

(see (1.4.1)) and the assumption that $((c_n(T))) \in l_{p,\infty}$, although it is useful only if $p < q'$.

Notes

2.1. An account of the spectral decomposition of compact operators in Hilbert spaces and the necessary background material may be found in any book on operator theory or functional analysis; see, e.g., [32] and [83].

2.2. The main results of this section come from [34] and [35]; for the representation of a map that follows from rapid decay of its Gelfand numbers see [41]. The relevance of X having a basis, and the intimate connections that exist between this and the property that the sequence (S_n) of linear projection operators is bounded, are analysed in [36]. In the notes at the end of the next chapter we shall discuss the notions of spectral operators and well-bounded operators which are relevant to the discussion in this section. These present other perspectives to the problem of representing operators acting in Banach spaces.

2.3. Knowledge of the embeddings of Sobolev spaces in Lebesgue spaces, as well as of their discrete equivalents, is of central importance for the type of application of the abstract results made in this section. Suitable references are [1] and [32]. It is also crucial to be able to determine the duality maps explicitly, as is the case for $L_p(\Omega)$ and $\overset{0}{W}{}^1_p(\Omega)(1 < p < \infty)$, for example.

2.4. Lusternik–Schnirelmann theory is only touched on in this section, and then simply to establish the coincidence of eigenvalues in the problems of subsections 2.4.1 and 2.4.2. The standard reference is [89]; of these volumes, III is the one devoted to variational problems. Other recommended references are [29] and [82].

The Hardy-type operator introduced in subsection 2.4.2 has been the object of extensive study over many years and much is known about its properties as a map between a wide assortment of spaces. For instance, in [33], Chapter 2, the operator T,

$$(Tf)(x) = v(x) \int_a^x u(t) f(t) dt,$$

is defined for $-\infty \le a < b \le \infty$, $p, q \in [1, \infty]$, and u and v such that, for any $X \in (a, b)$, $u \in L_{p'}(a, X)$ and $v \in L_q(X, b)$, thus allowing for singularities at the end points of (a, b). The boundedness and compactness of T as a map between $L_p(a, b)$ and $L_q(a, b)$ are analysed for the full range of values of p and q; two-sided estimates and an asymptotic formula are derived for its approximation numbers.

Chapter 3

Representation of Bounded Linear Operators

In this final chapter we make a start on the difficult problem of representing linear maps (between Banach spaces) that are merely bounded. The magnitude of the task is underscored by the fact that even in a Hilbert space context, really satisfactory results are available only for normal operators. Moreover, the methods used in the spectral analysis of self-adjoint or normal maps are totally different from those brought into play to handle compact maps between Banach spaces. Much is left to be done, but we hope that the tentative and incomplete material to be presented may be of some use in the construction of a satisfactory theory. As in the case of compact operators, the presence of nonlinear duality maps is a major source of difficulty, and the resulting problems have not yet been resolved to any great extent. It remains to be seen whether or not concepts and results from the classical Hilbert space theory, which is developed solely by linear means, provide the best guide when studying the Banach space theory. Of course they do indicate ways in which to proceed, but it is quite possible that, to some extent, the classical techniques may serve as distractions and that what is really needed is a radically different, fundamentally nonlinear, approach. We give below some preliminary work that may help to arrive at such an approach.

While additional assumptions may be made from time to time, the underlying supposition throughout is that X and Y are (real), reflexive Banach spaces that are, together with their duals, strictly convex; as before, J_X and J_Y will stand for duality maps with corresponding gauge functions μ_X and μ_Y, respectively; I_X will denote the identity map from X to itself and T is assumed to be a bounded linear map from X to Y. In view of Corollary 1.1.28, J_X^{-1} is the duality map from X^* to X with gauge function μ_X^{-1}. We emphasise that the assumption that X and Y are real was only made for convenience and has no influence on the results obtained.

3.1   An integral representation of points of X

We wish to introduce spaces reminiscent of the X_n that featured so prominently in the analysis of the last chapter when compact maps were considered. With this in mind, we set, for each $\lambda \geq 0$,

$$A'_\lambda := \{x \in X : \|Tx\|_Y \geq \lambda \|x\|_X\} \tag{3.1.1}$$

and

$$B'_\lambda := {}^0 \operatorname{sp} \{J_X A'_\lambda\} = \{x \in X : \langle x, J_X z\rangle_X = 0 \text{ for all } z \in A'_\lambda\}. \tag{3.1.2}$$

Thus $B'^0_\lambda = \overline{\operatorname{sp} \{J_X A'_\lambda\}}$.

Lemma 3.1.1. *Let* $\lambda \geq 0$. *Then*

(i) $$A'_\lambda \cap B'_\lambda = \{0\}; \tag{3.1.3}$$

(ii) $$\|Tx\|_Y < \lambda \|x\|_X \quad \text{for all } x \in B'_\lambda \backslash \{0\}; \tag{3.1.4}$$

(iii) $$\ker(T) \supset \cap_{\lambda>0} B'_\lambda. \tag{3.1.5}$$

Proof. (i) If $x \in A'_\lambda \cap B'_\lambda$, then $\|x\|_X \mu_X (\|x\|_X) = \langle x, J_X x\rangle_X = 0$: thus $x = 0$.
(ii) Suppose that $x \in B'_\lambda \backslash \{0\}$ and $\|Tx\|_Y \geq \lambda \|x\|_X$. Then $x \in A'_\lambda \cap B'_\lambda = \{0\}$, a contradiction.
(iii) Let $x \in \cap_{\lambda>0} B'_\lambda$, $x \neq 0$. Then for any $\lambda > 0$, $\|Tx\|_Y < \lambda \|x\|_X$. Hence $x \in \ker(T)$. $\qquad\square$

For each $\lambda \geq 0$ let P'_λ denote the projection of X onto B'_λ, so that $P'_\lambda x$ is the point of B'_λ nearest to x: this point is unique as X is strictly convex. Let

$$C'_\lambda := J_X^{-1} B'^0_\lambda.$$

Proposition 3.1.2. *Let* $\lambda \geq 0$. *Then* $P'_\lambda x = 0$ *for all* $x \in C'_\lambda$. *Moreover,* $C'_\lambda = (I_X - P'_\lambda)X$ *and*

$$X = B'_\lambda \uplus C'_\lambda. \tag{3.1.6}$$

Proof. For all $x \in X$, since $z := P'_\lambda x$ is the point in B'_λ nearest to x,

$$\|x - z\|_X = \inf_{t \in \mathbb{R}} \|x - z + ty\|_X, \quad \text{for all } y \in B'_\lambda.$$

Hence for all $y \in B'_\lambda$,

$$\frac{d}{dt} \|x - z + ty\|_X \upharpoonright_{t=0} = 0,$$

so that

$$\langle y, J_X(x - z)\rangle_X = 0. \tag{3.1.7}$$

If $x \in C'_\lambda$, then $J_X x \in B'^0_\lambda$, which implies that

$$\langle y, J_X x \rangle \rangle_X = 0 \text{ for all } y \in B'_\lambda.$$

Since z is the unique point in B'_λ which satisfies (3.1.7), it follows that $z = P'_\lambda x = 0$: thus $C'_\lambda \subset (I_X - P'_\lambda)X$. Now suppose that $w \in (I_X - P'_\lambda)X$, $w = (I_X - P'_\lambda)x$, say. From (3.1.7) we see that $\langle y, J_X w \rangle \rangle_X = 0$ for all $y \in B'_\lambda$, and hence $w \in C'_\lambda$, giving $C'_\lambda = (I_X - P'_\lambda)X$.

Finally, note that for any $x \in X$ we have $x = P'_\lambda x + (I_X - P'_\lambda)x$; and $B'_\lambda \cap C'_\lambda = \{0\}$. Moreover, $x \perp^j y$ for all $x \in C'_\lambda$ and $y \in B'_\lambda$. Thus (3.1.6) follows and the proof is complete. $\qquad\square$

After this preparation we turn to representation matters, and begin with the elements of X. If $0 \le \lambda < \mu$, then clearly $A'_\mu \subset A'_\lambda$ and $B'_\lambda \subset B'_\mu$. By (3.1.5), $\cap_{\lambda > 0} B'_\lambda = \{0\}$ if $\ker(T) = \{0\}$, and since $A'_\lambda = \{0\}$ if $\lambda > \|T\|$, we see from (3.1.2) that $\cup_{\lambda > 0} B'_\lambda = X$. We say that the family $(B'_\lambda)_{\lambda > 0}$ is *right-continuous* at λ if $B'_{\lambda+0} := \cap_{\lambda' > \lambda} B'_{\lambda'} = B'_\lambda$. The family (B'_λ) does not have this property in general, but $(B'_{\lambda+0})$ clearly does. In view of this and the need to have right-continuity in what follows, **we shall hereafter consider the closed subspaces $B_\lambda := B'_{\lambda+0}$ rather than B'_λ and the projections P_λ of X onto the subspaces B_λ. Therefore $(B_\lambda)_{\lambda > 0}$ is a family of closed, non-decreasing and right-continuous subspaces of X which is such that $\cap_{\lambda > 0} B_\lambda \subset \ker(T)$ and $\cup_{\lambda > 0} B_\lambda = X$.**

From Lemma 3.1.1 we have

Lemma 3.1.3. *Let $\lambda \ge 0$ and set $A_\lambda := \cup_{\lambda' > \lambda} A'_{\lambda'} = \{x \in X : \|Tx\|_Y > \lambda \|x\|_X\}$. Then*

(i) $$A_\lambda \cap B_\lambda = \{0\}; \tag{3.1.8}$$

(ii) $$\|Tx\|_Y \le \lambda \|x\|_X \quad \text{for all } x \in B_\lambda \setminus \{0\}; \tag{3.1.9}$$

(iii) $$\ker(T) \supset \cap_{\lambda > 0} B_\lambda. \tag{3.1.10}$$

The family of projections $(P_\lambda)_{\lambda > 0}$ has the properties given in the following lemma. We denote the identity and zero operators on X by I_X, 0_X, respectively and let $P_\lambda := 0_X$ for $\lambda < 0$ and $P_\lambda := I_X$ for $\lambda > \|T\|$.

Lemma 3.1.4.

(i) *If $\lambda \le \mu$,*

$$P_\lambda P_\mu = P_\mu P_\lambda = P_\lambda;$$

(ii) *P_λ is right-continuous on $(0, \|T\|]$: for all $x \in X$ and all $\lambda > 0$,*

$$P_{\lambda+0} x := \lim_{\varepsilon \to 0+} P_{\lambda+\varepsilon} x = P_\lambda x;$$

(iii) *$P_\lambda = 0_X$ for all $\lambda \le 0$; $P_\lambda = I_X$ for all $\lambda \ge \|T\|$;*

(iv) *if* $\cap_{\lambda>0}B_\lambda = \{0\}$, *then for all* $x \in X$, *as* $\lambda \to 0+$,

$$P_\lambda x \rightharpoonup 0 \quad and \quad \|(I_X - P_\lambda)x\|_X \to \|x\|_X; \qquad (3.1.11)$$

thus if X *is uniformly convex,*

$$\lim_{\lambda\to 0+} P_\lambda = 0_X$$

and $(P_\lambda)_{\lambda>0}$ *is right-continuous on* $[0, \|T\|]$;

(v) *for all* $\lambda < \mu$ *and all* $x \in X$,

$$\|x - P_\lambda x\|_X \geq \|x - P_\mu x\|_X; \qquad (3.1.12)$$

(vi) $C_\lambda := J_X^{-1}B_\lambda^0 = (I_X - P_\lambda)X$, *and* $C_\lambda \perp^j B_\lambda$ *for all* $\lambda > 0$.

Proof. The only property that needs proof is (iv), as with the exception of (vi), the others are obvious from the preceding remarks, and (vi) is a consequence of Proposition 3.1.2. To prove (iv) we use Lemma 2.2.11 with $L = \cap_{\lambda>0}B_\lambda$ and $N = \overline{\cup_{\lambda>0}B_\lambda^0}$. We therefore have that $L^0 = N$ and, for all $x \in X$,

$$\|x\|_{X/L} = \sup_{\lambda>0} \|x\|_{X/B_\lambda} = \lim_{\lambda\to 0} \|x\|_{X/B_\lambda}, \qquad (3.1.13)$$

since the subspaces B_λ decrease as λ decreases and hence the norms $\|x\|_{X\setminus B_\lambda}$ increase; cf. (2.2.39). As before we have omitted the canonical quotient maps from (3.1.13) in the interest of simplicity of appearance. Let P_L be the projection of X onto $L = \cap_{\lambda>0}B_\lambda$ and set $E_L = I_X - P_L$. We now let $(\lambda_k)_{k\in\mathbb{N}}$ be a strictly decreasing sequence in $(0, \|T\|]$ which converges to 0 and repeat the argument in the proof of Lemma 2.2.12. Then for all $x \in X$, $P_{\lambda_k} x \rightharpoonup P_L x$ and $\|x - P_{\lambda_k}x\|_X \to \|x - P_L x\|_X$ as $k \to \infty$; the weak convergence here is strong if X is uniformly convex. Since the subspaces B_λ decrease as λ decreases, (3.1.11) follows. $\qquad \square$

A family of projections $(P_\lambda)_{\lambda>0}$ satisfying (i)–(iv) is called a **resolution of the identity** or **partition of the identity**. If X is a Hilbert space it is usually called a **spectral family**.

With $E_\lambda := I - P_\lambda$, the family $(E_\lambda)_{\lambda>0}$ is right-continuous on $(0, \|T\|]$ and satisfies the following properties:

(i) $$E_\mu E_\lambda = E_{\max\{\mu,\lambda\}} \text{ and } E_\lambda^2 = E_\lambda;$$

(ii) for all $x \in X$ and $\lambda < \mu$,

$$\|E_\lambda x\|_X \geq \|E_\mu x\|_X;$$

(iii) $$E_\lambda = I_X \quad \text{for} \quad \lambda \leq 0; E_\lambda = 0_X \quad \text{for} \quad \lambda \geq \|T\|;$$

(iv) if $\cap_{\lambda>0}B_\lambda = \{0\}$, then for all $x \in X$,

$$\lim_{\lambda\to 0+} E_\lambda x = x.$$

In view of (3.1.12) and since $C_\lambda = (I_X - P_\lambda)X \perp^j B_\lambda$ and $B_\lambda = P_\lambda X$, we have that for all $x \in X$,

$$\phi(\lambda; x) := \langle x, J_X E_\lambda x \rangle_X = \langle E_\lambda x, J_X E_\lambda x \rangle_X = \mu_X (\|E_\lambda x\|_X) \|E_\lambda x\|_X. \quad (3.1.14)$$

It follows that for each $x \in X$, $-\phi(\cdot; x)$ is a non-decreasing, right-continuous function on $(0, \|T\|]$. We assign to $-\phi(\cdot; x)$ the function ν defined on subintervals of $[0, \|T\|]$ by the prescription that for each $y, z \in [0, \|T\|]$ with $y \le z$,

$$\nu([y, z]) = -\phi(z^+) + \phi(y^-), \quad \nu([y, z)) = -\phi(z^-) + \phi(y^-),$$
$$\nu((y, z]) = -\phi(z^+) + \phi(y^+), \quad \nu((y, z)) = -\phi(z^-) + \phi(y^+).$$

The function ν is a non-negative additive function of intervals and hence admits a unique extension to a non-negative Borel measure ν on $[0, \|T\|]$; see [79], Chapter 10. The Lebesgue–Stieltjes integral $- \int_{(0, \|T\|]} d\phi(\lambda; x)$ is defined to be $\int_{(0, \|T\|]} d\nu$; it coincides with the Riemann–Stieltjes integral. On setting $\nu_X(t) = t\mu_X(t)(t \ge 0)$, we have

$$
\begin{aligned}
- \int_{(0, \|T\|]} d\nu_X(\|E_\lambda x\|_X) &= \int_{(0, \|T\|]} d\nu \\
&= \nu((0, \|T\|]) = -\phi(\|T\|; x) + \phi(0+; x) \\
&= \nu_X(\|E_L x\|_X,
\end{aligned}
\quad (3.1.15)
$$

since $E_{\|T\|} = 0$ and $\lim_{\lambda \to 0+} E_\lambda x = E_L x = (I_X - P_L)x$, where P_L is the projection onto $\cap_{\lambda > 0} B_\lambda$. If $\ker(T) = \{0\}$ then $\cap_{\lambda > 0} B_\lambda = \{0\}$ by Lemma 3.1.3; hence $P_L = 0_X$ and

$$- \int_{(0, \|T\|]} d\nu_X(\|E_\lambda x\|_X) = \nu_X(\|x\|_X). \quad (3.1.16)$$

The resolution of the identity $(P_\lambda)_{\lambda > 0}$ generates operator-valued measures on suitable subspaces of $(0, \infty)$. For any semi-closed interval $I = (\mu, \lambda]$, we set

$$P(\mu, \lambda] = P_\lambda - P_\mu.$$

Then, with $I_1 = (\mu_1, \lambda_1], I_2 = (\mu_2, \lambda_2]$,

$$P(I_1)P(I_2) = P(I_1 \cap I_2),$$

so that $P(I_1)P(I_2) = 0$ for disjoint I_1, I_2. If Σ denotes the class of all subsets of $(0, \infty)$ of the form $S = \cup_{j=1}^n I_j$, with the $I_j = (\mu_j, \lambda_j]$ disjoint, then $P(S) = \sum_{j=1}^n P(I_j)$, this being independent of the way S is decomposed and hence well defined; $P(S)$ is called a **spectral measure** of S on Σ. If X is a Hilbert space, the measure $P(S)$ can be extended to the class Σ_1 of subsets which are countable unions of disjoint semi-closed intervals, and then to the class of all Borel subsets of $(0, \infty]$; see [90], Lemma 7.3 and Theorem 7.1.

Let $S = (\lambda_N, \lambda_0] = \bigcup_{i=1}^{N} (\lambda_{N-i+1}, \lambda_{N-i}]$, $0 < \lambda_N < \lambda_{N-1} < \cdots < \lambda_0 \leq \|T\|$. Then

$$\sum_{i=1}^{N} \left(P_{\lambda_{N-i}} - P_{\lambda_{N-i+1}} \right) = P(S) = P_{\lambda_0} - P_{\lambda_N}.$$

On choosing $\lambda_0 = \|T\|$ and allowing $\lambda_N \to 0$, it follows that if X is uniformly convex,

$$x - P_L x = \int_{(0,\|T\|)} dP_\lambda x \qquad (3.1.17)$$

for any $x \in X$. If also $\ker(T) = \{0\}$, then $P_L = 0_X$ and hence

$$x = \int_{(0,\|T\|)} dP_\lambda x, \qquad (3.1.18)$$

where the integral is the limit in X of the Riemann–Stieltjes sums.

We summarise these results in the following

Theorem 3.1.5. *Let* P_λ ($\lambda > 0$), P_L *denote the projections of* X *onto* B_λ, $\cap_{\lambda>0} B_\lambda$ *respectively, and set* $E_\lambda = I_X - P_\lambda$, $E_L = I_X - P_L$ *and* $\nu_X(t) = t\mu_X(t)$ ($t \geq 0$). *Then for all* $x \in X$,

$$-\int_{(0,\|T\|)} d\nu_X \left(\|E_\lambda x\|_X \right) = \nu_X \left(\|E_L x\|_X \right),$$

which becomes

$$-\int_{(0,\|T\|)} d\nu_X \left(\|E_\lambda x\|_X \right) = \nu_X \left(\|x\|_X \right)$$

if $\ker(T) = \{0\}$. *If* X *is uniformly convex and* $\ker(T) = \{0\}$, *then for all* $x \in X$,

$$x = \int_{(0,\|T\|)} dP_\lambda x.$$

Note that when X is a Hilbert space, we may take $J_X = I_X$, and then $B_\lambda^0 = B_\lambda^\perp$, while (3.1.6) becomes the orthogonal decomposition $X = B_\lambda \oplus B_\lambda^\perp$. Every $x \in X$ has the unique representation $x = x_\lambda + x_\lambda^\perp$, where $x_\lambda \in B_\lambda$, $x_\lambda^\perp \in B_\lambda^\perp$ and $P_\lambda x = x_\lambda$. Thus $E_\lambda x = x_\lambda^\perp$, and for all $y \in X$,

$$(E_\lambda x, y)_X = \left(x_\lambda^\perp, y_\lambda + y_\lambda^\perp \right)_X = \left(x_\lambda^\perp, y_\lambda^\perp \right)_X = \left(x, y_\lambda^\perp \right)_X = (x, E_\lambda y)_X.$$

Hence $E_\lambda^* = E_\lambda$; E_λ is therefore an orthogonal projection, having the properties $E_\lambda^* = E_\lambda$, $E_\lambda^2 = E_\lambda$. Furthermore, P_λ is an orthogonal projection, with $P_\lambda X$ and $(I_X - P_\lambda)X$ orthogonal complements in X.

3.2 An integral representation for T

The analysis of the last section gives a representation of Tx immediately when X is uniformly convex. For then we have from (3.1.17), since $P_L x \in \ker(T)$ by (3.1.10), that

$$Tx = \int_{(0,\|T\|]} dT\,P_\lambda x. \tag{3.2.1}$$

It would be desirable to have a representation in the form

$$Tx = \int_S \lambda dF_\lambda x$$

for some suitable integration set S and some natural operators F_λ, perhaps obtained by a Radon–Nikodym argument. However, we are only able to obtain a result of this nature in the special case when X is a Hilbert space and hence the family $(P_\lambda)_{\lambda>0}$ generates operator-valued measures on the Borel subsets of $[0,\|T\|]$. To be specific, for any fixed $x \in X$, set

$$\tau(\lambda) = (P_\lambda x, x)_X, \quad \kappa(\lambda) = \langle TP_\lambda x, J_Y Tx\rangle_Y,$$

and for any Borel set E in $(0,\|T\|]$,

$$\tau(E) = (P(E)x, x)_X, \quad \kappa(E) = \langle TP(E)x, J_Y Tx\rangle_Y.$$

We recall that the class of Borel subsets coincides with the set of countable unions $I = \cup_{j\in\mathbb{N}} I_j$ of disjoint sets I_j of the form $(a_j, b_j]$, and their complements $(0,\|T\|]\backslash I$. Then $P(I_j) = P_{b_j} - P_{a_j}$, $P(I) = \sum_{j\in\mathbb{N}} P(I_j)$ and $P((0,\|T\|] \setminus I) = I_X - P(I)$. We claim that $\tau(E) = 0$ implies that $\kappa(E) = 0$. For, since $P(E) = P^2(E)$ and $P^*(E) = P(E)$, it follows that

$$\tau(E) = (P(E)x, x)_X = \|P(E)x\|_X^2$$

and so $\tau(E) = 0$ is equivalent to $\|P(E)x\|_X = 0$. Hence

$$|\kappa(E)| \le \|T\|\|P(E)x\|_X\|J_Y Tx\|_{Y'} = 0.$$

Therefore, by the Radon–Nikodym theorem (see [83], Theorem 7.21-A), it follows that there exists $f(\cdot; x) \in L(0,\|T\|]$ such that

$$\langle Tx, J_Y Tx\rangle_Y = \int_{(0,\|T\|]} f(\lambda; x)d(P_\lambda x, x)_X. \tag{3.2.2}$$

We can also obtain a representation for T directly. Define the analogue of B_λ in the range of T by $\hat{B}_\lambda = T\widetilde{B}_\lambda$, where

$$\widetilde{B}_\lambda := \{x \in X : \langle Tx, J_Y Ty\rangle_Y = 0 \text{ for all } y \in A_\lambda\} \quad (\lambda > 0). \tag{3.2.3}$$

Thus $\hat{B}_\lambda = T(X) \cap {}^0(J_Y T(A_\lambda))$.

Lemma 3.2.1.
$$\cup_{\lambda>0}\left(A_\lambda \cap \widetilde{B}_\lambda\right) \subset \ker(T) \subset \cap_{\lambda>0}\widetilde{B}_\lambda. \tag{3.2.4}$$

Proof. Let $x \in A_\lambda \cap \widetilde{B}_\lambda$ for some $\lambda > 0$. Then

$$\mu_Y\left(\|Tx\|_Y\right)\|Tx\|_Y = \langle Tx, J_Y Tx\rangle_Y,$$

and hence $x \in \ker(T)$. If $x \in \ker(T)$, then $x \in \widetilde{B}_\lambda$ for all $\lambda > 0$. $\square$

Lemma 3.2.2. *Suppose that* $\ker(T) = \{0\}$. *Then* $\|Tx\|_Y / \|x\|_X \le \lambda$ *for all* $x \in$ $\widetilde{B}_\lambda \backslash \{0\}$ *and all* $\lambda > 0$.

Proof. If the claim were false, then $x \in A_\lambda$ for some $\lambda > 0$, so that by Lemma 3.2.1, $x \in A_\lambda \cap \widetilde{B}_\lambda = \{0\}$, a contradiction. $\square$

For each $\lambda > 0$ denote the projection of Y onto $\hat{B}_\lambda$ by Q_λ; let $Q_{\hat{L}}$ stand for the projection of Y onto $\hat{L} := \cap_{\lambda>0}\hat{B}_\lambda$. Since $\hat{B}_\lambda \subset \hat{B}_\mu$ if $0 < \lambda < \mu$, we then have

$$\|(I_Y - Q_\lambda)y\|_Y \ge \|(I_Y - Q_\mu)y\|_Y \text{ for all } y \in Y.$$

If $\lambda > \|T\|$, then $\widetilde{B}_\lambda = X$, $\hat{B}_\lambda = T(X)$ and $Q_\lambda Tx = Tx$ for all $x \in X$. The analysis leading to (3.1.16) and (3.1.17) now yields

$$\nu_Y\left(\left\|(I_Y - Q_{\hat{L}})Tx\right\|_Y\right) = \int_{(0,\|T\|)} d\nu_Y\left(\|(I_Y - Q_\lambda)Tx\|_Y\right) \quad (x \in X), \tag{3.2.5}$$

where $\nu_Y(t) = t\mu_Y(t)$, and when Y is uniformly convex, to

$$(I_Y - Q_{\hat{L}})Tx = \int_{(0,\|T\|)} dQ_\lambda Tx \ (x \in X). \tag{3.2.6}$$

This would be more aesthetically pleasing if $Q_{\hat{L}} = 0$, that is, if $\cap_{\lambda>0}\hat{B}_\lambda = \{0\}$. This is indeed the case in the corresponding situation when T is compact. For then, the projections that correspond to the Q_λ are the Q_n in Theorem 2.2.19, Q_n being the projection of Y onto the subspace Y_n. By (2.2.60), $\cap_{n\in\mathbb{N}}Y_n = \{0\}$ and hence Q_0, the analogue of $Q_{\hat{L}}$, is the zero operator.

Remark 3.2.3. It is worthwhile to pause at this point to consider the special case of Hilbert spaces X and Y. If $Y = X$, then to any spectral family of orthogonal projections which possesses the properties of the family $(P_\lambda)_{\lambda>0}$ in Section 3.2, there is associated a bounded, non-negative, self-adjoint operator $S : X \to X$ expressed by

$$S = \int_{(0,\|S\|)} \lambda dP_\lambda.$$

Conversely, any bounded, non-negative, self-adjoint operator $S : X \to X$ can be represented in this way. Moreover, S commutes with each projection P_λ. If X and Y are Hilbert spaces and $T : X \to Y$ is bounded, then $S := T^*T : X \to X$ is self-adjoint and has the above integral representation. Thus in the Hilbert space case, a dominant role is played by the map T^*T, and it is natural to look for analogues of this map in the general situation. One candidate is the nonlinear operator $S := J_X^{-1}T^*J_YT$ from X to X. This has the property that for all $\lambda > 0$, $\widetilde{B}_\lambda = {}^0(J_X S A_\lambda)$ since $z \in \widetilde{B}_\lambda$ if and only if

$$0 = \langle Tz, J_Y Tx\rangle_Y = \langle z, J_X Sx\rangle_X \text{ for all } x \in A_\lambda.$$

Also,

$$\langle x, J_X Sx\rangle_X = \langle Tx, J_Y Tx\rangle_Y = \mu_Y \left(\|Tx\|_Y\right) \|Tx\|_Y \text{ for all } x \in X$$

and if $Q_{\hat{L}}Tx = 0$,

$$(Sx, x)_X = \langle x, J_X Sx\rangle_X = \int_{(0,\|T\|]} d\nu_Y \left(\|(I_Y - Q_\lambda)Tx\|_Y\right)$$

where $\nu(t) = t\mu_Y(t)$. However, another map with strong claims is $J_X S = T^*J_Y T : X \to X^*$. (Note that both S and $J_X S$ reduce to T^*T when only Hilbert spaces are involved.) To analyse these, let $x, y \in X$. Then

$$\langle x - y, J_X Sx - J_X Sy\rangle_X = \mu_Y \left(\|Tx\|_Y\right) \|Tx\|_Y + \mu_Y \left(\|Ty\|_Y\right) \|Ty\|_Y$$
$$- \langle x, J_X Sy\rangle_X - \langle y, J_X Sx\rangle_X.$$

Since

$$|\langle x, J_X Sy\rangle_X| = |\langle Tx, J_Y Ty\rangle_X| \le \|Tx\|_Y \mu_Y \left(\|Ty\|_Y\right)$$

it follows that

$$\langle x - y, J_X Sx - J_X Sy\rangle_X \le \mu_Y \left(\|Tx\|_Y\right) \|Tx\|_Y + \mu_Y \left(\|Ty\|_Y\right) \|Ty\|_Y$$
$$- \|Tx\|_Y \mu_Y \left(\|Ty\|_Y\right) - \|Ty\|_Y \mu_Y \left(\|Tx\|_Y\right)$$
$$= \left(\mu_Y \left(\|Tx\|_Y\right) - \mu_Y \left(\|Ty\|_Y\right)\right) \left(\|Tx\|_Y - \|Ty\|_Y\right) \ge 0.$$

Hence $J_X S$ is monotone. Moreover,

$$\langle x, J_X Sx\rangle_X = \langle Tx, J_Y Tx\rangle_Y = \mu_Y \left(\|Tx\|_Y\right) \|Tx\|_Y \ge 0.$$

Thus if T has trivial kernel and positive minimum modulus $\gamma(T)$ (see [32], p. 28), so that $\|Tx\|_Y \ge \gamma(T) \|x\|_X$ for all $x \in X$,

$$\langle x, J_X Sx\rangle_X \ge \gamma(T) \|x\|_X \mu_Y \left(\gamma(T) \|x\|_X\right),$$

which implies that $J_X S$ is coercive. If (x_n) is a sequence in X such that $x_n \to x \in X$, then by Proposition 1.1.26, $J_X Sx_n \to J_X Sx$. Thus by Theorem 1.1.27, $J_X S$ is surjective. We note additionally that if $\gamma(T) \ge 1$ and $\mu_X = \mu_Y = \mu$, then

$$\langle x, J_X Sx\rangle_X \ge \gamma(T) \|x\|_X \mu \left(\|x\|_X\right) = \gamma(T) \langle x, J_X x\rangle_X.$$

Thus
$$\langle x, J_X Sx \rangle_X - \lambda \langle x, J_X x \rangle_X \geq (\gamma(T) - \lambda) \langle x, J_X x \rangle_X ,$$
which can be written in terms of the semi-inner product on X as
$$(Sx, x)_X - \lambda (x, x)_X \geq (\gamma(T) - \lambda) (x, x)_X .$$
If $\lambda < \gamma(T)$ and $\gamma(T) \geq 1$, we have
$$\| J_X Sx - \lambda J_X x \|_{X^*} \geq (\gamma(T) - \lambda) \| J_X x \|_{X^*} ,$$
from which it follows that the map $M : x \longmapsto J_X Sx - \lambda J_X x$ of X to X^* is injective.

We see that both S and $J_X S$ have interesting properties. However, the fact that both are, in general, nonlinear is a substantial stumbling block so far as integral representations are concerned. It is, of course, quite possible that it is a mistake to be influenced too much by the Hilbert space theory, the siren songs of which may simply lure us onto the rocks. Perhaps some radically different approach is necessary.

3.3 Compact operators revisited

Suppose that X is uniformly convex and let $T : X \to Y$ be compact, with j-eigenvectors (x_i) and j-eigenvalues (λ_i), $\ker(T) = \{0\}$ and $\operatorname{rank} T = \infty$. From Proposition 2.2.4 each j-eigenvalue is of finite multiplicity:
$$m(i) := \sharp\{x_j : \lambda_j = \lambda_i\} < \infty. \tag{3.3.1}$$
Denote the distinct j-eigenvalues by $\lambda'_j, j \in \mathbb{N}$, and define
$$\mathcal{A}_k := \bigcup_{i=1}^{k} \{x_j : \lambda_j = \lambda'_i\} \tag{3.3.2}$$
$$\mathcal{B}_n := {}^0 \operatorname{sp} \{J_X(\mathcal{A}_{n-1})\} = X_{s(n)}, \tag{3.3.3}$$
where $s(n) = \sum_{i=1}^{n} m(i)$. Also, for $\lambda'_n \leq \lambda < \lambda'_{n-1}$, let $\mathcal{B}_\lambda := \mathcal{B}_{\lambda'_n} = \mathcal{B}_n$ and denote the projection of X onto $\mathcal{B}_\lambda$ by $\mathcal{P}_\lambda$; recall that $(\lambda'_n)_{n \in \mathbb{N}}$ is decreasing. Then the subspaces $\mathcal{B}_\lambda$ are decreasing with λ and right-continuous; hence so is the family $(\mathcal{P}_\lambda)_{\lambda>0}$. Moreover, $\bigcap_{\lambda>0} \mathcal{B}_\lambda = \ker(T)$ by Proposition 2.2.16, and if $\lambda < \mu$,
$$\mathcal{P}_\lambda \mathcal{P}_\mu = \mathcal{P}_\mu \mathcal{P}_\lambda = \mathcal{P}_\lambda. \tag{3.3.4}$$

If $\lambda > \|T\|$ then $\mathcal{A}_\lambda = \varnothing$, $\mathcal{B}_\lambda = X$, $\mathcal{P}_\lambda = I_X$, and we have, for any $x \in X$, that $\lim_{\lambda \to 0} \mathcal{P}_\lambda x = \mathcal{P}_L x$, where $\mathcal{P}_L$ is the projection onto $\ker(T)$. We can now repeat the argument leading to Theorem 3.1.5 and obtain
$$x = \int_{(0, \|T\|]} d\mathcal{P}_\lambda x, \quad \text{for all } x \in X. \tag{3.3.5}$$

By construction $\mathcal{P}_\lambda$ is constant in each interval $[\lambda'_n, \lambda'_{n-1})$ and hence

$$x = \sum_{n=1}^\infty \left(\mathcal{P}_{\lambda'_n} - \mathcal{P}_{\lambda'_{n+1}}\right) x. \tag{3.3.6}$$

If X is a Hilbert space, then $J_X = I_X$ and from Remark 2.2.28, $\mathcal{P}_{\lambda'_n} = I_X - S_{s(n)}$, where $S_{s(n)}$ is the linear projection of X onto $Z_{s(n)-1} = \mathrm{sp}\{x_1, x_2, \ldots, x_{s(n)-1}\}$. In this case $\mathcal{P}_{\lambda'_n} - \mathcal{P}_{\lambda'_{n+1}} = S_{s(n+1)} - S_{s(n)}$ is a projection operator on the orthogonal complement of $Z_{s(n)-1}$ in $Z_{s(n+1)-1}$ (see [3], Section 37, (4°)), this being $\mathrm{sp}\{x_{s(n)}, \ldots, x_{s(n+1)-1}\}$. Hence

$$\left(\mathcal{P}_{\lambda'_n} - \mathcal{P}_{\lambda'_{n+1}}\right) x = \sum_{i=s(n)}^{s(n+1)-1} c_i(x)x_i, \quad x = \sum_{i=1}^\infty c_i(x)x_i,$$

and $c_i(x) = (x, x_i)_X$ since the x_i are orthonormal.

For a general Banach space X, all we seem to know is that, from (2.2.41),

$$\left(I_X - \mathcal{P}_{\lambda'_n}\right) x \in J_X^{-1} M_{s(n)-1} = J_X^{-1} \mathrm{sp}\left\{J_X x_1, \ldots, J_X x_{s(n)-1}\right\} =: K_{s(n)-1}, \tag{3.3.7}$$

say. Clearly $\mathrm{sp}\, K_{s(n)-1}$ contains $Z_{s(n)-1}$ and by Lemma 2.2.29 it is finite dimensional. In proving that $x = \lim_{n\to\infty}(I_X - \mathcal{P}_{\lambda_n})S_n x$, Theorem 2.2.14 is therefore of particular interest as it doesn't obviously follow from the integral representation (3.3.5).

Suppose now that we have the optimal situation (guaranteed by the boundedness of $(S_n)_{n\in\mathbb{N}}$), that for all $x \in X$,

$$x = \sum_{i=1}^\infty \xi_i(x)x_i = \sum_{n=1}^\infty \sum_{\lambda_i=\lambda_n} \xi_i(x)x_i, \tag{3.3.8}$$

where the $\xi_i(x)$ are defined in (2.2.18). For $\lambda'_n \leq \lambda < \lambda'_{n-1}$, define $S_\lambda = S_{s(n)}$ and $X_\lambda = X_{s(n)}$. Then $(S_\lambda)_{\lambda>0}$ is right-continuous, $I_X - S_\lambda$ is a linear projection onto X_λ and $S_\lambda S_\mu = S_{\min\{\lambda,\mu\}}$. From the identity

$$S_{\lambda'_N} x - S_{\lambda'_0} x = \sum_{i=0}^{N-1} \left(S_{\lambda'_{i+1}} - S_{\lambda'_i}\right) x$$

and the observation that the choice $\lambda'_0 > \|T\|$ yields $X_{\lambda'_0} = X$ and $S_{\lambda'_0} = 0_X$, we have

$$x = \lim_{N\to\infty} S_{\lambda'_N} x = \int_{(0,\|T\|]} dS_\lambda x, \quad \text{for all } x \in X. \tag{3.3.9}$$

Let R_k be the operator in (2.2.53), that is,

$$R_k y = \sum_{i=1}^{k-1} \gamma_i(y)y_i, \quad y_i = Tx_i/\lambda_i,$$

where the coefficients $\gamma_i(y)$ are given in (2.2.54). Define $R_\lambda := R_{\lambda'_n} = R_{s(n)}$ for $\lambda'_n \leq \lambda < \lambda'_{n-1}$. Then R_λ is a linear projection of TX onto TX_λ, $(R_\lambda)_{\lambda > 0}$ is right-continuous and by (2.2.57), $R_\lambda Tx = TS_\lambda x$ for all $x \in X$. It now follows (even if $\ker(T) \neq \{0\}$) that

$$Tx = \int_{(0,\|T\|]} dR_\lambda Tx, \quad \text{for all } x \in X. \tag{3.3.10}$$

Notes

Other contributions to the problems discussed in the last two chapters have been based on the imposition of additional conditions on the operator T and not on the spaces X and Y. A notable example is the body of work on spectral operators described by Dunford and Schwarz in [31]. We give a brief sketch of this, but refer the reader to [31] and the references therein, for a comprehensive account; [26] is also recommended, especially as it includes a treatment of the well-bounded operators mentioned below. A bounded linear operator $T : X \to X$ (X a general complex Banach space) is defined to be a **spectral operator** if there exists $E : \mathcal{B} \to B(X)$ (where $\mathcal{B}$ is the family of all Borel subsets of $\mathbb{C}$) such that for each $\delta \in \mathcal{B}$ the map $E(\delta)$ is a projection and the following conditions hold:

1. $(E(\cdot))$ is countably additive in the strong operator topology, i.e., for every sequence (δ_n) of disjoint sets in $\mathcal{B}$ and every $x \in X$,

$$E\left(\cup_{n \in \mathbb{N}} \delta_n\right) x = \sum_{n=1}^{\infty} E(\delta_n)x;$$

2. $TE(\delta) = E(\delta)T$, $\delta \in \mathcal{B}$;
3. for any $\delta \in \mathcal{B}$, the spectrum of the restriction of T to $E(\delta)(X)$ lies in $\bar{\delta}$;
4. $E(\emptyset) = 0$, $E(\mathbb{C}) = I$, $E(\mathbb{C} \setminus \delta) = I \setminus E(\delta)$;
5. $E(\delta \cap \mu) = E(\delta) \wedge E(\mu) := E(\delta)E(\mu)$;
6. $E(\delta \cup \mu) = E(\delta) \vee E(\mu) := E(\delta) + E(\mu) - E(\delta)E(\mu)$;
7. $\sup_{\delta \in \mathcal{B}} \|E(\delta)\| < \infty$.

The family $(E(\cdot))$ is a resolution of the identity in the sense of section 3.2 but has additional properties involving its interaction with T. It is called the **spectral resolution**, or **resolution of the identity** of T; it is unique. A main result is that a bounded operator T is a spectral operator if and only if it is the sum $T = S + N$ of a bounded **spectral-type** operator S given by

$$S = \int_{\sigma(S)} \lambda E(d\lambda)$$

in terms of a resolution of the identity $E(\cdot)$ defined on the spectrum $\sigma(S)$ of S, and a quasi-nilpotent operator N (i.e., its spectral radius is zero) which commutes with S. The decomposition is unique. Moreover, T and S have the same resolution of the identity and spectrum. An operational calculus for these spectral operators is then developed. If the spectrum $\sigma(T)$ of T is countable (for instance if T is compact), then any $x \in X$ has an unconditionally convergent expansion of the type

$$x = \sum_n x_n = \sum_{\lambda \in \sigma(T)} E(\lambda)x,$$

where x_n is a kind of "generalized eigenvector" associated with $\lambda_n \in \sigma(T)$. If X is reflexive, $\sigma(T) \subseteq [a,b]$ for some $a, b \in \mathbb{R}$ and

$$\|p(T)\| \leq K \sup_{t \in [a,b]} |p(t)|$$

for all polynomials p, then the generalized eigenvectors x_n are eigenvectors in the ordinary sense: $(T - \lambda_n I)x_n = 0$. For finite-dimensional spaces, the spectral resolution of T is its canonical Jordan reduction in classical matrix theory; thus in a finite-dimensional space X, every linear operator is a spectral operator. In Hilbert spaces, every bounded self-adjoint or normal operator is a spectral operator on account of the spectral theorem. It is shown that the theory of normal operators is not an infallible guide even to the theory of arbitrary operators in finite-dimensional spaces, and other considerations are necessary for a general reduction theory.

In [81], Smart defined an operator $T : X \to X$ to be **well-bounded** if there exists a positive real constant K and a compact interval $[a, b]$ such that for all polynomials p,

$$\|p(T)\| \leq K \left(|p(b)| + \int_a^b |p'(t)| dt \right).$$

He then proved in [81] that if T is well bounded and X is reflexive, there exists for any $t \in \mathbb{R}$, a unique projection $E(t)$ in $B(X)$ which is such that

1. $E(t)$ commutes with any operator which commutes with T;
2. $\|E(t)\| \leq 2K \quad (t \in \mathbb{R})$;
3. $E(t) = 0 \quad$ for $\quad t < a \quad$ and $\quad E(t) = I \quad$ for $\quad t \geq b$;
4. $E(s) = E(s)E(t) = E(t)E(s) \quad$ for $\quad s \leq t$;
5. $\lim_{t \to s+} E(t)x = E(s)x \quad (x \in X, s \in \mathbb{R})$;
6. $\sigma\left(T \upharpoonright_{E(t)X}\right) \subseteq (-\infty, t] \cap \sigma(T) \quad$ and $\quad \sigma\left(T \upharpoonright_{[I-E(t)]X}\right) \subseteq [t, \infty) \cap \sigma(T)$, for $t \in \mathbb{R}$.

Furthermore,

$$T - \int_{a+}^b t\, dE(t)$$

is quasinilpotent. It was subsequently proved by Ringrose in [77] that this operator is in fact 0, so

$$T = \int_{a+}^{b} t dE(t).$$

In [90], Zarantonello makes a general study of projections on convex sets in Hilbert spaces, motivated by the objective to establish a spectral integration theory with respect to measures generated by a family of commuting projection operators (a resolution of the identity). The resulting integrals are of the form

$$T = \int_{-\infty}^{\infty} \lambda dP_\lambda;$$

these are operators acting in the Hilbert space, nonlinear in general, which are shown to generalise linear self-adjoint operators whose properties they mimic. He establishes in [90], Theorem 10.3 necessary and sufficient conditions for an operator T to be expressible as a spectral integral

$$T = \int_{0}^{\infty} \lambda dP_\lambda$$

in terms of the resolution of the identity $(P_\lambda)_{\lambda>0}$. For linear operators T the criteria require T to be non-negative and self-adjoint. The theorem also allows T to be non-linear, and although this is of theoretical interest, Zarantonello was not convinced of its value in deciding if a given operator T is of the displayed form.

Zarantonello makes the interesting observation that while the classification of real points in the spectrum of a linear operator T can either be given by means of the properties of the resolution of the identity (P_λ), or by the existence and properties of the resolvent operator $(T - \lambda I)^{-1}$, this is not so if T is not linear. To quote Zarantonello: "It appears as if the operation of taking the inverse of $(T - \lambda I)$ is too tied up to linearity, and that its consideration for nonlinear operators is somewhat artificial as it imposes on them unnatural linear constraints."

Bibliography

[1] Adams, R.A., *Sobolev spaces*, Academic Press, New York, 1975.

[2] Abramowitz, M. and Stegun, I.A., *Handbook of mathematical functions*, Dover, New York, 1973.

[3] Akhiezer, N.I. and Glazman, I.M., *Theory of linear operators in Hilbert space*, volume I, Pitman, Boston-London-Melbourne, 1981.

[4] Alber, Ya.I., James orthogonality and orthogonal decompositions in Banach spaces, J. Math. Anal. Appl. **312** (2005), 330–342.

[5] Ball, K., Carlen, E.A. and Lieb, E.H., Sharp uniform convexity and smoothness inequalities for trace norms, Invent. Math. **115** (1994), 463–482.

[6] Banach, S., *Théorie des opérations linéaires*, 2nd revised edition, Chelsea, New York, 1999.

[7] Bennewitz, C. Approximation numbers = singular values, J. Comp. and App. Math., **208** (2005), 102–110.

[8] Binding, P., Boulton, L., Čepička, J., Drábek, P. and Girg, P., Basis properties of eigenfunctions of the p-Laplacian, Proc. American Math. Soc. **134** (2006), 3487–3494.

[9] Binding, P. and Drábek, P., Sturm–Liouville theory for the p-Laplacian, Studia Sci. Math. Hungar. **40** (2003), 375–396.

[10] Binding, P.A. and Rynne, B.P., The spectrum of the periodic p-Laplacian, J. Diff. Eq., **235** (2007), 199–218.

[11] Bourbaki, N., *Espaces vectoriels topologiques*, Hermann, Paris, 1956.

[12] Browder, F.E., Nonlinear operators and nonlinear equations of evolution in Banach spaces, Proc. AMS Symp. Pure Math., Vol. XVIII, Part 2, American Math. Soc., Providence, Rhode Island, 1976.

[13] Brown, B.M. and Eastham, M.S.P., Eigenvalues of the radial p-Laplacian with a potential on $(0, \infty)$, J. Comp. and App. Math., **208** (2006), 111–119.

[14] Brown, B.M. and Eastham, M.S.P., Bounds for the positive eigenvalues of the p-Laplacian with decaying eigenvalues, J. Comp. and App. Math., **325** (2007), 734–744.

[15] Brown, B.M. and Reichel, W., Eigenvalues of the radially symmetric p-Laplacian in $\mathbb{R}^n$, J. London Math. Soc. (2) **69** (2004), 657–675.

[16] Bushell, P.J. and Edmunds, D.E., Remarks on generalised trigonometric functions, Rocky Mountain J. Math. **42** (2012), 25–57.

[17] Carl, B., Entropy numbers, s-numbers, and eigenvalue problems, J. Funct. Anal. **41** (1981), 290–306.

[18] Carl, B., Entropy numbers of nuclear operators, Studia Math. **77** (1983), 155–162.

[19] Carl, B., On s-numbers, quasi s-numbers, s-moduli and Weyl inequalities of operators in Banach spaces, Rev. Math. Complut. **23** (2010), 467–487.

[20] Carl, B., Hinrichs, A. and Rudolph, P., Entropy numbers of convex hulls in Banach spaces and applications, preprint.

[21] Carl, B. and Stephani, I., *Entropy, compactness and the approximation of operators*, Cambridge Univ. Press, Cambridge, 1990.

[22] Carl, B. and Triebel, H., Inequalities between eigenvalues, entropy numbers, and related quantities of compact operators in Banach spaces, Math. Ann. **251** (1980), 129–133.

[23] Clarkson, J.A., Uniformly convex spaces, Trans. American Math. Soc. **40** (1936), 396–414.

[24] Davies, E.B., Non-self-adjoint differential operators, Bull. London Math. Soc., **34** (2002), 513–532.

[25] Day, M.M., Some more uniformly convex spaces, Bull. American Math. Soc. **47** (1941), 504–507.

[26] Dowson, H.R., *Spectral theory of linear operators*, Academic Press, London, 1978.

[27] Drábek, P., Kufner, A. and Nicolesi, F., *Nonlinear elliptic equations*, Univ. West Bohemia, Pilsen, 1996.

[28] Drábek, P. and Manásevich, R., On the solution to some p-Laplacian nonhomogeneous eigenvalue problems, Diff. and Int. Eqns. **12** (1999), 723–740.

[29] Drábek, P. and Milota, J., *Methods of nonlinear analysis; applications to differential equations*, Birkhäuser, Basel-Boston-Berlin, 2007.

[30] Drábek, P. and Takác, P., On variational eigenvalues of the p-Laplacian which are not of Ljusternik-Schnirelmann type, J. London Math. Soc., (2) **81** (2010), 625–649.

[31] Dunford, N. and Schwartz, J.T., *Linear operators, Part III; Spectral operators*, Wiley-Interscience, 1971.

[32] Edmunds, D.E. and Evans, W.D., *Spectral theory and differential operators*, Oxford University Press, Oxford, 1987.

[33] Edmunds, D.E. and Evans, W.D., *Hardy operators, function spaces and embeddings*, Springer, Berlin-Heidelberg-New York, 2004.

[34] Edmunds, D.E., Evans, W.D. and Harris, D.J., Representations of compact linear operators in Banach spaces, J. London Math. Soc. (2) **78** (2008), 65–84.

[35] Edmunds, D.E., Evans, W.D. and Harris, D.J., A spectral analysis of compact linear operators in Banach spaces, Bull. London Math. Soc. **42** (2010), 726–734; Corrigendum, Bull. London Math. Soc. **44** (2012), 1079–1081.

[36] Edmunds, D.E., Evans, W.D. and Harris, D.J., The structure of compact linear operators in Banach spaces, Rev. Mat. Complut. **26** (2) (2013), 445–469.

[37] Edmunds, D.E., Gurka, P. and Lang, J., Properties of generalized trigonometric functions, J. Approx. Th. **164** (2012), 47–56.

[38] Edmunds, D.E. and Lang, J., The j-eigenfunctions and s-numbers, Math. Nachr. **283** (2010), 463–477.

[39] Edmunds, D.E. and Lang, J., *Eigenvalues, embeddings and generalised trigonometric functions*, Lecture Notes in Mathematics 2016, Springer, Heidelberg-Dordrecht-London-New York, 2011.

[40] Edmunds, D.E. and Lang, J., Gelfand numbers and widths, J. Approx. Th. **166** (2013), 78–84.

[41] Edmunds, D.E. and Lang, J., Explicit representation of compact linear operators in Banach spaces via polar sets, Studia Math. **214** (2013), 78–84.

[42] Edmunds, D.E. and Triebel, H., *Function spaces, entropy numbers, differential operators*, Cambridge Univ. Press, Cambridge, 1996.

[43] Edwards, R.E., *Fourier series, A modern introduction*, volume 2, Springer-Verlag, Berlin-Heidelberg-New York, 1982.

[44] Elbert, Á., A half-linear second order differential equation, in *Qualitative theory of differential equations*, volumes I,II, (Szeged, 1979), pp. 153–180, Colloq. Math. Soc. János Bolyai, 30, North-Holland, Amsterdam-New York, 1981.

[45] Enflo, P., A counterexample to the approximation in Banach spaces, Acta Math. **130** (1973), 309–317.

[46] Fabian, M., Habala, P., Santalucía, V.M., Pelant, J. and Zizler, V., *Functional analysis and infinite-dimensional geometry*, Springer-Verlag, Berlin-Heidelberg-New York, 2001.

[47] Friedlander, L., Asymptotic behavior of the eigenvalues of the p-Laplacian, Comm. Partial Diff. Equations **14** (1989), 1059–1069.

[48] Garcia Azorero, J.P. and Peral Alonso, I., Comportement asymptotique des valeurs propres du p-laplacien, C. R. Acad. Sci. Paris **301** (1988), 75–78.

[49] Gilbarg, D. and Trudinger, N.S., *Elliptic partial differential equations of second order*, Grundlehren der mathematischen Wissenshaften 224, Springer-Verlag, Berlin-Heidelberg-New York, 1977.

[50] Gurarii, V.I. and Gurarii, N.I., On bases in uniformly convex and uniformly smooth spaces, Izv. Akad. Nauk SSSR Ser. Mat. **35** (1971), 210–215.

[51] Hanner, O., On the uniform convexity of L^p and l^p, Ark. Math. **3** (1956), 239–244.

[52] James, R.C., Orthogonality and linear functionals in normed linear spaces, Trans. American Math. Soc. **61** (1947), 265–292.

[53] James, R.C., Weak compactness and reflexivity, Israel J. Math. **2** (1964), 101–119.

[54] James, R.C., Weakly compact sets, Trans. American Math. Soc. **113** (1964), 129–140.

[55] James, R.C., Characterizations of reflexivity, Studia Math. **23** (1964), 205–216.

[56] James, R.C., Superreflexive spaces with bases, Pacific J. Math. **41-42** (1972), 409–419.

[57] Karlin, S., Bases in Banach spaces, Duke Math. J. **15** (1948), 971–985.

[58] Kato, T., *Perturbation theory for linear operators*, 2nd edition, Springer-Verlag, Berlin-Heidelberg-New York, 1976.

[59] König, H., *Eigenvalue distribution of compact operators*, Birkhäuser, Basel, 1986.

[60] Kufner, A., John, O. and Fučik, S., *Function spaces*, Academia, Prague, 1977.

[61] Le, An, Eigenvalue problems for the p-Laplacian, Nonlinear Anal. **64** (2006), 1057–1099.

[62] Lindenstrauss, J. and Tzafriri, L., *Classical Banach spaces* I and II, Springer-Verlag, Berlin-Heidelberg-New York, 1977 and 1979.

[63] Lindqvist, P., Some remarkable sine and cosine functions, Ricerche Mat. **44** (1995), 269–290.

[64] Lindqvist, P. and Peetre, J., *p*-arclength of the *q*-circle, Math. Student **72** (2003), no. 1-4, 139–145.

[65] Martin, R.H., *Nonlinear operators and differential equations in Banach spaces*, Wiley, New York, 1976.

[66] McArthur, C.W., Developments in Schauder basis theory, Bull. American Math. Soc. **78** (1972), 877–907.

[67] Milman, D., On some criteria for the regularity of spaces of the type (B), Comptes Rendus de l'Académie des Sciences de l'URSS **20** (1938), 243–246.

[68] Morrison, T.J., *Functional Analysis: introduction to Banach space theory*, Wiley, New York, 2001.

[69] Nörlander, G., The modulus of convexity in normed linear spaces, Arkiv. Mat. **4** (1960), 15–17.

[70] Oja, E., A short proof of a characterization of reflexivity of James, Proc. American Math. Soc. **126** (1998), 2507–2508.

[71] Peetre, J., The differential equation $y'^p - y^p = \pm 1$ $(p > 0)$, Ricerche Mat. **43** (1994), 91–128.

[72] Pietsch, A., *Operator ideals*, North-Holland, 1980.

[73] Pietsch, A., *Eigenvalues and s-numbers*, Cambridge Univ. Press, Cambridge, 1987.

[74] Pietsch, A., *History of Banach spaces and linear operators*, Birkhäuser, Boston-Basel-Berlin, 2007.

[75] Pietsch, A., Bad properties of the Bernstein numbers, Studia Math. **184** (2008), 263–269.

[76] Pinkus, A., *n-widths in approximation theory*, Springer-Verlag, Berlin, 1985.

[77] Ringrose, R., On well-bounded operators, J. Austral. Math. Soc. **1** (1960), 334–343.

[78] Robertson, A.P. and Robertson, W.J., *Topological Vector Spaces*, Cambridge University Press, Cambridge, 1964.

[79] Rudin, W., *Principles of Mathematical Analysis*, 2nd edition, McGraw-Hill, New York, 1964.

[80] Ryan, R.A., *Introduction to tensor products of Banach spaces*, Springer, London-Berlin-Heidelberg, 2002.

[81] Smart, D.R., Conditionally convergent spectral expansions, J. Australian Math. Soc. Ser. A **1** (1960), 319–333.

[82] Struwe, M., *Variational methods*, Springer-Verlag, Berlin-Heidelberg-New York, 1990.

[83] Taylor, A.E., *Introduction to functional analysis*, Wiley, New York, 1958.

[84] Titchmarsh, E.C., *Eigenfunction expansions associated with second-order differential equations*, Part 1, 2nd edition, Oxford University Press, Oxford, 1958.

[85] Triebel, H., *Interpolation theory, function spaces, differential operators*, North-Holland, Amsterdam, 1978

[86] Whittaker, E.T. and Watson, G.N., *Modern analysis*, Cambridge University Press, Cambridge, 1952.

[87] Wojtaszczyk, P., *Banach spaces for analysts*, Cambridge University Press, Cambridge, 1991.

[88] Yosida, K., *Functional analysis*, Springer-Verlag, Berlin-Göttingen-Heidelberg, 1965.

[89] Zeidler, E., *Nonlinear functional analysis and its applications*, volumes I, IIA, IIB, III and IV, Springer-Verlag, Berlin-Heidelberg-New York, 1986.

[90] Zarantonello, E.H., Projections on convex sets in Hilbert spaces and spectral theory, pp. 237–424, in *Contributions to nonlinear functional analysis*, (Proceedings of a Symposium held at the Math. Research Center, University of Wisconsin, Madison, Wisconsin, 1971) Academic Press, 1971.

Author Index

Abramowitz, M., 44
Adams, R.A., 126
Alber, Ya.I., 21

Ball, K., 6, 7
Banach, S., 65
Bennewitz, C., 118, 119
Binding, P., 46, 113, 116, 117
Boulton, L., 46
Bourbaki, N., 8, 72, 93
Browder, F.E., 17
Brown, B.M., 111, 112
Bushell, P.J., 46, 51, 65

Carl, B., 58, 59, 66, 125
Carlen, E.A., 6, 7
Čepička, J., 46
Clarkson, J.A., 6

Davies, E.B., x
Day, M.M., 5
Dowson, H.R., 138
Dràbek, P., 46, 109, 113, 116, 117, 126
Dunford, N., x, 138

Eastham, M.S.P., 112
Edmunds, D.E., vii, 6, 46, 51, 58, 61, 65, 77, 83, 100, 103, 105, 107, 108, 120, 121, 126
Edwards, R.E., 24
Elbert, Á., 113
Enflo, P., 40, 90
Evans, W.D., vii, 58, 105, 107, 108, 126

Fabian, M., 65, 98
Friedlander, L., 107

Garcia Azorero, J.P., 107
Gilbarg, D., 112
Girg, P., 46
Gurarii, N.I., 65
Gurarii, V.I., 65
Gurka, P., 65

Habala, P., 65, 98
Harris, D.J., vii, x, 126

James, R.C., ix, 9, 19, 20, 65, 77

Karlin, S., 65
Kato, T., 34
König, H., 59, 107, 124, 125
Kufner, A., 109

Lang, J., vii, x, 6, 61, 65, 77, 83, 100, 103, 120, 121
Le, An, 116, 121
Lieb, E.H., 6, 7
Lindenstrauss, J., 5, 8, 40, 65
Lindqvist, P., 65

Martin, R.H., 83
McArthur, C.W., 65
Milman, D., 8
Milota, J., 126
Morrison, T.J., 65

Nicolesi, F., 109
Nörlander, G., 15

Oja, E., 65

Peetre, J., 65
Pelant, J., 65, 98
Peral Alonso, I., 107
Pietsch, A., 40, 57, 58, 65, 66, 104

Pinkus, A., 66

Reichel, W., 111
Ringrose, R., 140
Robertson, A.P., 99
Robertson, W.J., 99
Rudin, W., 131
Ryan, R.A., 40
Rynne, B.P., 117

Santalucía, V.M., 65, 98
Schwartz, J.T., x, 138
Smart, D.R., 139
Stegun, I.A., 44
Stephani, I., 58, 66
Struve, M., 126

Takàc, P., 117
Taylor, A.E., 93, 126, 133
Titchmarsh, E.C., 111
Triebel, H., 25, 26, 58, 59
Trudinger, N.S., 112
Tzafriri, L., 5, 8, 40, 65

Watson, G.N., 53
Whittaker, E.T., 53

Zarantonello, E.H., 140
Zeidler, E., 126
Zizler, V., 65, 98

Subject Index

absolute value, 68
additive s-function, 56
Alber's theorem, 21
approximation numbers $a_n(S)$, 57
approximation property (AP), 37
Arzelà–Ascoli theorem, 118

ball-measure of noncompactness, 58
Banach–Steinhaus theorem, 32
basic sequence, 21
basis, 21
basis constant, 24
Bernstein numbers $b_n(S)$, 57
Bessaga–Pełczyński selection
    principle, 28, 92
Beta function, 43
biorthogonal functionals, 29
biorthogonal system, 29
Birkhoff, 20
Borel measure, 131
boundedly complete basis, 30
Brouwer's fixed point theorem, 17

canonical map, 72
compact, vii, 1
compact self-adjoint operator, 67

Dirichlet eigenvalue problem, 106
duality map, 15

Emden–Fowler equation, 107
entropy numbers $e_n(S)$, 58

Fréchet derivative, 10
Fréchet-differentiable, 10
Friedrichs inequality, 105

Gâteaux derivative, 10

Gâteaux-differentiable, ix, 10
gauge function, 15
Gelfand numbers $c_n(S)$, 57, 60
Gelfand widths $\tilde{c}_n(S)$, 60
generalised trigonometric functions,
    42
gradient, 10
Gram–Schmidt type procedure, 94

Haar functions, 24
Hahn–Banach theorem, 2, 8, 12, 15,
    40
Hardy-type operator, 118
Hausdorff
    –Young theorem, 123
    distance, 61
    metric, 64
Hilbert numbers $h_n(S)$, 57
hypergeometric function, 44

incomplete Beta function, 44
injective s-function, 57

j-eigenfunctions, ix
j-eigenvalues, ix, 77
j-eigenvectors, 77
j-orthogonal, 20
James, 20
James orthogonal direct sum, 21

Kolmogorov numbers $d_n(S)$, 57
Krasnoselski genus, 115

Lebesgue spaces, 6
Lebesgue–Stieltjes integral, 131
linear projections S_k, 79
linear span, 1

Lusternik–Schnirelmann, ix, 115,
    121

Mazur, 27
metric injection, 56
metric surjection, 57
modulus of convexity, 3
modulus of smoothness, 13
monotone basis, 24
monotone map, 16
multiplicative s-function, 56

nonlinear projections P_k, 81
norm, 1
normal operators, x
normalised, 21
normalised basis, 35
nuclear map, 123

p-biharmonic operator, ix, 112
p-Laplacian, ix, 105
p-trigonometric functions, 42
partition of the identity, 130
polar set, 20
precompact, 1
projection, 9
projective limit, 99
Prüfer-type transformation, x

quasi-nilpotent operator, 139

r-nuclear map, 123
r-nuclear norm, 123
Radon–Nikodym theorem, 133
resolution of the identity, x, 130
Riemann–Stieltjes integral, 131
Riesz basis, 24

s-numbers, 56
Schatten p-class, 124
Schauder basis, 21
semi-inner product, 19
seminormalised basis, 35
shrinking basis, 30
singular values, 68
Sobolev embedding theorem, 108
spectral,
    -type, 138
    family, 130
    measure, 131
    operator, 138
strictly convex, vii, 2
Sturm–Liouville, ix
supporting functional, 11
surjective s-function, 57

uniformly
    convex, ix, 3
    Fréchet-differentiable, 13
    smooth, ix, 13

weak convergence, 1
well-bounded, 139
Weyl inequalities, 59
Weyl numbers $w_n(S)$, 57

Notation Index

A_λ, 129
A'_λ, 128
$a_n(S)$, 57

$B(a,b)$, 43
B_λ, 129
$\hat{B}_\lambda$, 133
B'_λ, 128
$b_n(S)$, 57
$B(X)$, 1
B_X, 1
$B(X,Y)$, 1

$C_{1/p'}$, 65
C'_λ, 128
$\widetilde{c}_n(S)$, 60
$c_n(S)$, 57
$\cos_{p,q}$, 52
$\cos y_p$, 43

$\delta(A,B)$, 61
$\dim X$, 1
$\widetilde{d}(M,N)$, 61
$d(M,N)$, 61
$d_n(S)$, 57
Δ_p, 106
$\delta_p(\varepsilon)$, 8
$\delta_X(\varepsilon)$, 3

$e_n(S)$, 58

$\gamma_j(y)$, 88
$\operatorname{grad} \|x\|$, 11

$h_n(S)$, 57

$I(\cdot;a,b)$, 44

J, 15
J_X, 73
$\tilde{J}_X$, 72

μ, 15
$M_1 \uplus M_2$, 21
$M_2 \perp^j M_1$, 21
M_k, 75

N_k, 75
ν_X, 132

P_K, 9
P_k, 81
P_λ, 129
P_N, 21
π_p, 42
$\pi_{p,q}$, 52

R_k, 88
$\rho_X(\tau)$, 13

$S_{1/p'}$, 65
$\sin_p$, 43
$\sin_p^{-1}$, 44
$\sin_{p,q}$, 52
S_k, 79
$s_n(S)$, 56
$\operatorname{sp} S$, 1
S_X, 1

$\|T\|$, 1
$\tan_p$, 44

$u_{(p)}$, 113

$\overset{0}{W}{}_p^1(\Omega)$, 105
$w_n(S)$, 57
$W_p^1(\Omega)$, 105

X, vii
$(x, h)_X$, 19
$\langle x, x^* \rangle_X$, 1

$\xi_j(x)$, 76
$x \perp^j y$, 20
X_k, 75

Y_k, 75

Z_{k-1}, 79

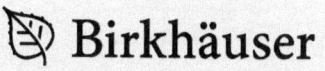 **Birkhäuser** | **www.birkhauser-science.com**

Operator Theory: Advances and Applications (OT)

This series is devoted to the publication of current research in operator theory, with particular emphasis on applications to classical analysis and the theory of integral equations, as well as to numerical analysis, mathematical physics and mathematical methods in electrical engineering.

Edited by
Joseph A. Ball (Blacksburg, VA, USA), Harry Dym (Rehovot, Israel),
Marinus A. Kaashoek (Amsterdam, The Netherlands), Heinz Langer (Vienna, Austria),
Christiane Tretter (Bern, Switzerland)

■ **OT 237: Kaashoek, M.A. / Rodman, L. / Woerdeman, H.J.** (Eds.), Advances in Structured Operator Theory and Related Areas. The Leonid Lerer Anniversary Volume (2013).
ISBN 978-3-0348-0638-1

■ **OT 236: Cepedello Boiso, M. / Hedenmalm, H. / Kaashoek, M.A. / Montes Rodríguez, A. / Treil, S.** (Eds.), Concrete Operators, Spectral Theory, Operators in Harmonic Analysis and Approximation. 22nd International Workshop in Operator Theory and its Applications, Sevilla, July 2011 (IWOTA11) (2013).
ISBN 978-3-0348-0647-3

■ **OT 234/OT235: Eidelman, Y. / Gohberg, I. / Haimovici, I.** (Eds.), Separable Type Representations of Matrices and Fast Algorithms.

Vol. 1. Basics. Completion problems. Multiplication and inversion algorithms (2013).
ISBN 978-3-0348-0605-3

Vol 2. Eigenvalue method (2013).
ISBN 978-3-0348-0611-4

■ **OT 233: Todorov, I.G. / Turowska, L.** (Eds.), Algebraic Methods in Functional Analysis. The Victor Shulman Anniversary Volume (2013).
ISBN 978-3-0348-0501-8

■ **OT 232: Demuth, M. / Kirsch, W.** (Eds.), Mathematical Physics, Spectral Theory and Stochastic Analysis (2013).
ISBN 978-3-0348-0590-2

■ **OT 231: Molahajloo, S. / Pilipović, S. / Toft, J. / Wong, M.W.** (Eds.), Pseudo-Differential Operators, Generalized Functions and Asymptotics (2013).
ISBN 978-3-0348-0584-1

■ **OT 230: Brown, B.M. / Eastham, M.S.P. / Schmidt, K.M.**, Periodic Differential Operators (2013).
ISBN 978-3-0348-0527-8

■ **OT 229: Almeida, A. / Castro, L. / Speck, F.-O.** (Eds.), Advances in Harmonic Analysis and Operator Theory. The Stefan Samko Anniversary Volume (2013).
ISBN 978-3-0348-0515-5

■ **OT 228: Karlovich, Y.I. / Rodino, L. / Silbermann, B. / Spitkovsky, I.M.** (Eds.), Operator Theory, Pseudo-Differential Equations, and Mathematical Physics (2013).
ISBN 978-3-0348-0536-0

■ **OT 227: Janas, J. / Kurasov, P. / Laptev, A. / Naboko, S.** (Eds.), Operator Methods in Mathematical Physics. Conference on Operator Theory, Analysis and Mathematical Physics (OTAMP) 2010, Bedlewo, Poland (2013).
ISBN 978-3-0348-0530-8

■ **OT 226: Alpay, D. / Kirstein, B.**, Interpolation, Schur Functions and Moment Problems II (2012).
ISBN 978-3-0348-0427-1

■ **OT 225: Sakhnovich, L. A.**, Levy Processes, Integral Equations, Statistical Physics: Connections and Interactions (2012).
ISBN 978-3-0348-0355-7

■ **OT 224: Benguria, R. / Friedman, E. / Mantoiu, M.** (Eds.), Spectral Analysis of Quantum Hamiltonians. Spectral Days 2010 (2012).
ISBN 978-3-0348-0413-4